Professional Communications

Other Titles of Interest

Civil Engineering Practice in the Twenty-First Century:
Knowledge and Skills for Design and Management,
Neil S. Grigg, Marvin E. Criswell, Darrell G. Fontane, and Thomas J. Stiller.
2001. ISBN 0-7844-0526-3.

Engineering Your Future, Second Edition: The Non-Technical Side
of Professional Practice in Engineering and Other Technical Fields,
Stuart G. Walesh.
2000. ISBN 0-7844-0489-5.

Excellent Communication Skills Required for Engineering Managers,
Todd A. Shimoda.
1994. ISBN 0-7844-0047-4.

How to Produce Effective Operations and Maintenance Manuals,
Mike Tidwell.
2000. ISBN 0-7844-0011-3.

How to Work Effectively with Consulting Engineers:
Getting the Best Project at the Right Price
(ASCE Manuals and Reports on Engineering Practice No. 45, revised edition). 2003.
ISBN 0-7844-0637-5.

Managing and Leading: 52 Lessons Learned for Engineers,
Stuart G. Walesh.
2004. ISBN 0-7844-0675-8.

Personal Success Strategies: Developing Your Potential.
1999. ISBN 0-7844-0446-1.

Professional Communications

A Handbook for Civil Engineers

Heather Silyn-Roberts, Ph.D.

Library of Congress Cataloging-in-Publication Data

Silyn-Roberts, Heather.
   Professional communications : a handbook for civil engineers / Heather Silyn-Roberts.
     p. cm.
   Includes bibliographical references and index.
   ISBN 0-7844-0732-0
     1. Civil engineers. 2. Communication. I. Title

   TA157.S49 2004
   624'.01'4—dc22   2004056801

Published by American Society of Civil Engineers
1801 Alexander Bell Drive
Reston, Virginia 20191
www.asce pubs.asce.org

Any statements expressed in these materials are those of the individual authors and do not necessarily represent the views of ASCE, which takes no responsibility for any statement made herein. No reference made in this publication to any specific method, product, process or service constitutes or implies an endorsement, recommendation, or warranty thereof by ASCE. The materials are for general information only and do not represent a standard of ASCE, nor are they intended as a reference in purchase specifications, contracts, regulations, statutes, or any other legal document. ASCE makes no representation or warranty of any kind, whether express or implied, concerning the accuracy, completeness, suitability, or utility of any information, apparatus, product, or process discussed in this publication, and assumes no liability therefore. This information should not be used without first securing competent advice with respect to its suitability for any general or specific application. Anyone utilizing this information assumes all liability arising from such use, including but not limited to infringement of any patent or patents.

ASCE and American Society of Civil Engineers—Registered in U.S. Patent and Trademark Office.

Photocopies: Authorization to photocopy material for internal or personal use under circumstances not falling within the fair use provisions of the Copyright Act is granted by ASCE to libraries and other users registered with the Copyright Clearance Center (CCC) Transactional Reporting Service, provided that the base fee of $25.00 per article is paid directly to CCC, 222 Rosewood Drive, Danvers, MA 01923. The identification for this book is 0-7844-0732-0/05/$25.00. Requests for special permission or bulk copying should be addressed to Permissions & Copyright Dept., ASCE.

13 12 11   5 6 7

Contents

Preface and Introduction

This is a book I have wanted to write for a long time. I have worked as a documentation consultant to large engineering organizations for 15 years and have run many programs to help people write effectively. I do not need to convince engineers that it is important. Most of you know how vital it is in the global engineering workplace to write documents that have one basic aim: *to convey the writer's information to the reader accurately and simply.*

Yet my experiences with engineers in North America, Europe, and Australasia show that there is a widespread problem. Many of you have told me that you can find it difficult to extract information from documents, even those written by an engineer of the same discipline. Other professionals in the engineering workplace, such as financial, legal, and planning personnel, tell me of their difficulties reading engineering documents. Many major organizations are concerned about the professionalism of their documentation and its effect on their competitiveness.

Written material is one of the major outputs of a professional engineer, and career development can be enhanced or hindered by one's competence. In today's professional environment, many engineers spend a considerable amount of time in writing, either in the primary process itself or in monitoring and correcting other people's work. They also say that the demand for it is increasing over the years.

When it is done well, writing can be a satisfying part of an engineer's work. All that is needed are exactly those characteristics that make a good engineer: the ability to gather facts and present them logically in a way that satisfies the recipients' needs. Yet many of you, at ease with mathematical and design processes, tell me you feel uncomfortable about writing. And many of you chose to become engineers because you realized that your talents lay in the technical area rather than in creating a flow of words.

My goal for this book is for it to be relentlessly practical. It is for those of you who would like a handbook giving concise information about what is needed in an engineering document. It takes into account your skills and your specific difficulties. I know them well.

I have also taken into account all your preferences for the layout of a book. A great number of you have told me that you do not have the time—or the inclination—to read long passages of text. Most of you prefer concise, listed material to long paragraphs. Indeed you may not even have read this far into the preface since it is in the form of solid text. The rest of the book is different.

I am excited to have written this book. It is based on real-world experience with practicing engineers. I hope it helps you.

Organization of this book

The book has been designed as a user-centered tool to allow you to access information readily. It has a somewhat unusual format. It is set out as a modular reference handbook that does not need to be read sequentially.

Part 1 deals with the primary design and organization of the information in a document. It concentrates on the needs of the reader, as opposed to how the writer would like to present the information. There is a world of difference between these two viewpoints.

In many ineffective reports, it is often not the wording itself that is faulty but rather the placement and flow of information. It could be easy to assume that, since technical documents need to be subdivided into appropriate sections, the flow of information will arise as a matter of course. This is not so. Even in a heavily sectioned document the placement of the information can be ineffective. Information structure is critical because, to retrieve material from a document, engineers go through specific reading processes. If the structure thwarts those processes, then the reader can waste a lot of time trying to understand it.

Part 2 consists of one large chapter that gives guidelines on how to write many of the sections that a document may possibly need. Most of these sections are described in terms of the purpose of the section, how to approach writing it, the difficulties associated with it, and mistakes that are commonly made.

Part 3 deals with specific types of documents that are commonly written by civil engineers, from complex documents such as large proposals and feasibility studies to simple faxes, e-mails, and memos.

Part 4 presents material on referencing, revising, proofreading, and reviewing, together with the commonly used editorial conventions.

Part 5 deals with problems of style and how to correct them. Many engineers are aware that they often cannot recognize when their style is incorrect. For this reason, this book concentrates on common stylistic errors and how to recognize and correct them. This approach flies in the face of current educational principles, which emphasize the need to present only the correct form. However, I have found that engineers react positively to the approach taken in this chapter. It barely nods at classic grammar.

Part 6 gives guidelines on how to present work orally. The verbal presentation of technical material is tightly bound up with the writing process. Most people go through a form of writing before making an oral presentation, and the structure of the information is as significant as it is in a written document. There is also the crucial point that many engineers are concerned about their performance and level of nervousness. Making an oral presentation is a critical component of an engineer's professional life; that is the reason it's in the book.

How to use this book

To understand the basic elements of a well-organized and well-presented document:

Chapter 1: *The Basics of Technical Writing* provides the basics of concise technical writing.

Chapter 2: *The Structure of an Engineering Document* gives guidelines about how to design information to make it accessible to a reader.

Chapter 3: *Organizing a Document and Choosing Appropriate Sections* describes the basic skeleton of sections, choosing other sections, planning a document, and presentation style.

Chapter 4: *Presentation Style* gives guidelines for page layout, page orientation, font, type size, and binding.

To write a specific type of document:

Use the Table of Contents to find the chapter that deals with that specific type of document (**Chapters 6–13**). Use it in conjunction with **Chapter 5,** *Requirement for Sections and Elements of a Document*; this gives the requirements for the most commonly used sections in many types of reports.

To write an Executive Summary, Summary, or Abstract:

Use **Chapter 6,** *Summarizing.*

Other chapters that will help with the process are in Section 4:

Chapter 14: *Referencing Your Sources*

Chapter 15: *Editorial Conventions*

Chapter 16: *Revising, Proofreading, and Reviewing*

To check on writing style and how to correct common problems:

Use **Chapter 17,** *Problems of Style: Recognizing and Correcting Common Mistakes.*

To present work orally:

> Use **Chapters** **18** and **19**. Chapter 18 covers presentation at a seminar or conference. Chapter 19 deals with presenting to a small group.

Acknowledgments

Many people have helped me with the content of this book. I would particularly like to thank engineers at Maunsell and URS for invaluable advice. Interaction with Babcock and Beca International Consultants was also fruitful. In addition, my thanks are due to the many professional engineering colleagues and academics in the United States, the United Kingdom, Germany, and Australasia with whom I have discussed these issues. I am also grateful to the thousands of professional engineers and students I have taught, who in turn have sharpened my thoughts about these issues. Staff at the University of Auckland's Engineering Library have been very helpful. Members of the editorial staff at ASCE Press have given me excellent direction. To all of them I offer the usual protection: the familiar rubric that the book's mistakes are solely my own responsibility.

About the Author

Heather Silyn-Roberts earned her B.Sc. at the University of Wales, United Kingdom, and her Ph.D. at the University of Auckland, New Zealand. She was also a post-doctoral fellow in Biomechanics at the University of Tübingen. Dr. Silyn-Roberts is currently a Senior Lecturer in Mechanical Engineering at the University of Auckland. She has published two books, *Writing for Science and Engineering: Papers, Presentations and Reports* (2000, Butterworth-Heinemann) and *Writing for Science: a Practical Handbook for Science, Engineering and Technology Students*, 2nd Edition (2002, Pearson Education).

International Best Practice in Report Writing: Getting Started

Part 1 is designed to give an introduction to best practice in engineering writing.

It starts by describing the characteristic features of effective and ineffective documents. It gives solutions to the problems, poses questions that need to be answered, and provides an action plan for writing an effective document.

It then describes the basic principles of structuring information in a document so that readers can readily access it. This needs an understanding of how busy professional engineers extract information from a document.

It presents a sequence of steps needed to produce a well-organized document, together with a list of the possible sections for many types of documents and their brief requirements.

It finishes by giving general indications of page layout, orientation, font, and binding, while acknowledging that most documents are governed by an organization's house style.

1

The Basics of Technical Writing

This chapter is designed to give an introduction to international best practice in engineering writing. It points out the characteristic features of good documents. It also outlines the ineffective features, those that engineers frequently complain about. It then provides solutions to the problems: questions that need to be asked and an action plan for writing an effective document.

Structure of the chapter

1.1   What to aim for: characteristics of an effective document

Most technical writing calls for exactly the characteristics of a good engineer: the ability to gather facts and present them logically in a way that satisfies the recipients' needs. Do not be concerned that the output is in written form: once you have gone through the initial logical processes, all that is left is to express it as simply and clearly as possible. Nothing else.

The main characteristics to aim for are:

• **A document that is obviously aimed at the readers' needs.**
This calls for an examination by you of what the reader needs. It means moving away from what you think you need to write and determining from the

readers' points of view what you believe they need to receive. There is a world of difference between the two.

• **Clear main points and an obviously logical flow of information.**
This needs clear thinking, clear identification of the main points, clear information structure, and effective formatting. These should be the major goals when producing written material.

• **Clearly written Executive Summary, Conclusions, and Recommendations.**
These critical sections are often read in isolation by readers who need to determine the main points. Moreover, the conclusions and recommendations of a report are usually the chief outcomes of a document. They need to be both clear and well supported.

• **Concise, clear prose.**
We are not aiming here for literary style. Your style needs only to be concise, simple, clear, and grammatically correct.

• **Illustrations that are well designed—it should be obvious why each has been included.**
We are visual animals; we love illustrations and often look at them early in the reading process in isolation from the main text. Make your figures and tables professional.

• **Formatting that is well conceived and that helps the understanding of the document.**
Formatting of a high standard is a major feature of a professional document.

1.2   Pitfalls to avoid

Recognizing the undesirable characteristics of ineffective documents will help you avoid them. It also will save you time.

The following points are engineers' chief complaints about the documents they need to read.

• **A document in which the main points are not clear; a poorly planned document; a document with ineffective structure and organization.**
These are the most common complaints, since the reader needs to work hard to extract the information.

• **Inadequately supported conclusions.**
Since the conclusions are critical, they cannot be taken as valid if they are not supported.

• **Recommendations that are not clear, either in intent or in wording.**
The recommendations in a report are the stimulus for action. If they are not clear, the document loses much of its validity.

- **Documents that are too long and wordy.**
There is often an unconscious need to impress either a client or a superior. This can make people mistakenly use long words and complicated constructions, and include too much unnecessary material, in the belief that the document looks impressive. Reports should aim to inform, not to impress. In the end, it is clarity that impresses, not showiness.

- **Too many unfamiliar and inadequately defined acronyms and abbreviations.**
This has become a major complaint. Acronyms are a feature of professional life that are designed to save time and space. But frustration runs very high when they are not understood or are ill-defined.

- **An Executive Summary that is uninformative.**
The Executive Summary is a critical section of a document. It does not serve its purpose if it omits vital information.

- **Style and grammar that reflect a lack of basic grammar and writing skills.**
Many engineers realize somewhat belatedly that their skills in this area are inadequate. The only remedy is to take steps to improve. It is going to be vital for your future career.

- **Too much solid text with a lack of visual interest.**
Readers can feel daunted and become bored when presented with large amounts of solid text. Text should be broken up; however, the following point also needs to be observed.

- **Overfragmentation of text into bulleted lists (the opposite of the previous point).**
Bullets are useful for efficient transmission of the main points. However, text should not be overfragmented into chaos.

- **Excessively broken-up or otherwise inappropriate formatting.**
The formatting capacities of word processors are powerful. There is a temptation to overuse them. Formatting should have only one purpose: to aid the understanding of the document.

1.3   Solutions to the main problems: questions and action plan

Effective solutions combine a rigorous determination of what the readers need to hear from you and an understanding of how your document will be read.

1.3.1   Questions

To avoid the main problems, we need to answer two questions:

• Who will the readers be?

The usual advice is: Identify your reader and write specifically for him or her. But engineers know that they rarely write for one individual or a group of individuals with equivalent knowledge and expertise. Moreover, a report can be passed on to an unexpected readership. There certainly is no time to write a number of different reports directed at different readerships. One report therefore needs to be understood by all readers.

• What will readers need to help them skim and extract information?

The answer: a report that is carefully planned and logically structured, which also has an easily identifiable pathway through it. Chapters 2 and 3 of this book show how to plan a document so that the information is readily accessible and how to construct a pathway by using strategically placed overview information.

1.3.2   Action plan for writing an effective document

Here is an action plan for writing a report that has clear main points, is easy to navigate through, and has text that is well structured and in an acceptable style.

1. Spend time planning the structure of the document.

Busy engineers tend not to take enough time in planning a document. But by doing so, the overall writing time is shortened.

2. Identify the main point(s) of the document from the point of view of the readers, not from your own point of view. Make sure that the main points are obvious to the readers.

Place yourself in the readers' minds and work out what questions they will most want answered. Do not write from the point of view of "What do I want to write?" Instead ask yourself, "What do the readers need to hear from me?"

3. Place overviews throughout the document.

Initial overview information is essential for readily assessing the whole document. Brains instinctively look for it. Strategically placed overviews help all readers—of whatever level of knowledge—to understand the document.

> ▸ Place a fully informative Summary or Executive Summary at the beginning of all engineering documents (with the possible exception of some specifications), giving an overview of the whole document.
> ▸ Use section summaries to give a brief overview of the findings or content of each section.
> ▸ Use topic sentences for paragraphs to give an overview of the material in that paragraph.

4. Design the document around the basic skeleton of sections.

Documents have the same basic structure: They start and end with the same sections.

Title
Summary/Executive Summary
Introduction/Background
⋮
(middle sections)
⋮
Conclusions
Recommendations *(if needed) or* Conclusions and Recommendations
Appendixes *(if needed)*

5. Add to the basic skeleton of sections as needed.

A document is constructed by using the basic skeleton and adding sections as required by the document.

6. Make sure that the sections and the flow of information in them are logically ordered.

Using the Outline mode of a word processor can help achieve this.

7. Make sure that each of the conclusions in the Conclusions section is fully supported by the material in the rest of the document.

Unsupported conclusions reduce the validity of the document.

8. Make sure that any recommendations arise directly from the conclusions.

9. Use a Glossary of Terms and Abbreviations.

A glossary defines the terms and abbreviations that you think the reader will need.

10. Write clearly and concisely.

If you have doubts about your grammar and writing style, it would be best to do something about it. Poor word use, grammar, and spelling make a document appear unprofessional.

- ▶ Do not use incomplete sentences.
- ▶ Use verbs correctly.
- ▶ Use punctuation effectively.
- ▶ Use the correct form for plurals.
- ▶ Use the correct word or pairs of words.
- ▶ Avoid prepacked jargon and clichés.
- ▶ Make your text clear and concise: Avoid long sentences, big words, and complex constructions.

2

The Structure of an Engineering Document

This chapter describes the basic principles of the structure of information in a document so that it can be readily retrieved by its readers. It might easily be thought that readers' needs are served by the traditional subdivision of a technical document into sections. However, readers need more than this. An understanding of how busy professional engineers extract information from a document shows us how that information should be structured. This chapter builds its principles around this process.

Structure of the chapter

2.1 The traditional basic skeleton of most reports
2.2 A navigational pathway: the sections that engineers read first
2.3 Traditional report structure: the diamond structure of a document
2.4 Structure for an executive audience: nontraditional report structure
2.5 Sections of a document: also diamond-shaped
2.6 Helping nonengineers to understand a complex document
2.7 Deliberate repetition of information in a document

2.1   The traditional basic skeleton of most reports

The common basic structure for most types and lengths of professional documents is:

Title
Summary/Executive Summary

Introduction/Background

⋮

(middle sections)

⋮

Conclusions
Recommendations *(if needed) or* Conclusions and Recommendations
Appendixes *(if needed)*

There is a very good reason for this structure: It aids the readers' assessment and understanding of the document.

2.2   A navigational pathway: the sections that engineers read first

In the course of my work, I have asked over 1,000 engineers: What processes do you use to read and extract information from a document? The answers have been very illuminating.

The sequence in which a document is commonly read is:

1. The Title.
2. The Executive Summary. All engineers say that if the document has no summary, then the reading process is more difficult.
3. The Conclusions and Recommendations.
4. The required detail. (The Introduction is sometimes read before the Conclusions and Recommendations.)

This is the classic rule of thumb for obtaining an overview of a document: Title, Executive Summary, Introduction, Conclusions, and Recommendations. These sections form a navigational pathway of overview information. *Our brains need these overviews.* In reading a document this way, we are unconsciously doing what psychological studies have shown: that people assess complex information much better if they are given an undetailed overview of it first.

2.3   Traditional report structure: the diamond structure of a document

It is useful to think of the whole document as a diamond resting on a solid base (Figure 2-1).

The *diamond* is a diamond of information, its width reflecting the level of detail. At the two narrow ends, the information is brief, focused, and concise.

The *base* represents the solid detail of the Appendixes.

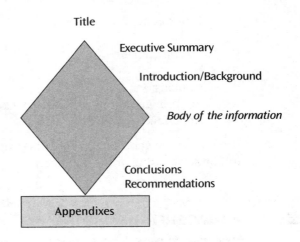

Figure 2-1   *Diagrammatic representation of the traditional structure of a document. The level of detail is low at the two narrow ends—the initial and final overview information (Title; Abstract, Summary, or Executive Summary; and Conclusions and Recommendations).*

We already know that to get a good overview of the whole document, most engineers first read the sections at the two narrow ends of the diamond.

Guidelines for information at the narrow ends of the document

Our brains are helped by overviews. The sections in a document that give overview information are the Title, the Executive Summary, section summaries, Conclusions, and Recommendations.

- The Title should adequately describe the contents of your document in the fewest possible words. It is the first overview presented to the readers; it should give them immediate access to the main subject matter.
- The Executive Summary should give an overview of the *whole* document: the context of the study, the means of doing it, the outcomes, the conclusions you have drawn, and the recommended actions (if any). It should be informative, not descriptive.
- The Introduction and/or Background should give the introductory material necessary to understand the document.
- The Conclusions section presents the deductions that you have made.

If recommendations are necessary, the Recommendations section gives the recommended actions that naturally follow from the conclusions. Note the following, however:

- The Recommendations can often be usefully placed immediately after the Executive Summary.
- The combined section Conclusions and Recommendations is often effective.

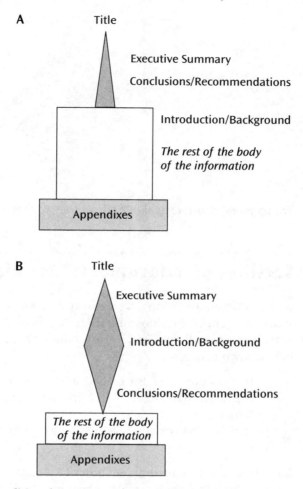

Figure 2-2   *Alternative, nontraditional report structures, aimed at executive audiences.*

Armed with the broad information in these sections, the reader is now prepared to assess the detailed information in the broad part of the diamond.

2.4   Structure for an executive audience: nontraditional report structure

Since many executive readers do not read the whole report, the Conclusions and Recommendations may be brought to the front, so that they follow the Executive Summary (Figure 2-2A). Alternatively, they may be placed immediately after the Introduction. This allows the summary and background material of the report to be read in isolation, without needing to read the detailed matter. The effect is one of a smaller diamond sitting on a stepped base (Figure 2-2B).

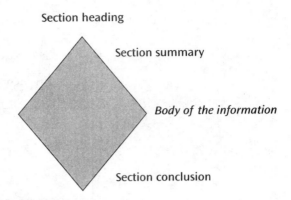

Figure 2-3   *The diamond structure applied to document sections.*

2.5   Sections of a document: also diamond-shaped

If your document is long, you can help a reader by applying the diamond shape to sections of a document as well as to the whole document (Figure 2-3). This is less obvious and not so commonly observed. However, it is a great help for two reasons:

- It gives nonspecialist readers a navigational pathway through a document that will help them to understand a complex document in broad terms.
- It also helps the specialist reader to better assess the complex information.

Guidelines for structuring a section of a document

▸ Use an informative section heading.

▸ Immediately below the section heading, place a very brief summary of the section (in nontechnical terms if possible). Title it Section Summary. If there is little content to include, describe what the section covers. Section summaries become very important in long reports to which various personnel contribute substantially different sections (e.g., structural material, architectural, environmental, and so on).

▸ If possible, end the section with a very brief conclusion to that section that will lead into the next section.

2.6   Helping nonengineers to understand a complex document

All readers benefit from a clear navigational pathway through a document. Specialist readers will automatically perceive the pathway of Executive Summary, Conclusions, Recommendations, and the section summaries. Nonexpert readers—for example, financial and legal management, local government authorities—often complain that engineers' documents are impenetrable. To help such readers, indicate that they can obtain an undetailed overview of the document by reading these specific sections.

Suggested wording (placed immediately before or after the main Summary):

> For overview information about this document, please read the Executive Summary, Conclusions, and Recommendations, together with the section summaries at the beginning of each section. *(Give page numbers if appropriate.)*

2.7   Deliberate repetition of information in a document

This section describes how information is deliberately repeated in the various sections of the basic skeleton.

People are sometimes concerned because they see information repeated throughout a report. But it has to be remembered that this repetition is deliberate and controlled—the basic diamond structure demands it. The repeated information forms part of the navigational pathway described previously and guides the reader through the document.

Table 2-1 is a guide to deliberate repetition of information throughout a document.

This deliberate restatement of undetailed information in the basic skeleton is a feature of a professional document. However, information that is repeated because the document has been sloppily assembled is another matter.

Conclusions

- Readers first need to read overview information to help them understand a document.
- A navigational pathway made up of overview material is provided by the Title, Executive Summary, Conclusions, Recommendations, and the section summaries. Most readers will read these sections before reading the detailed material.
- The structure of a document can be thought of as a diamond resting on a solid base.
- The diamond structure can also be used to structure sections of a document: Section heading; section summary; Conclusion (if appropriate).

Table 2-1. *Guide to the deliberate repetition of information in a document*

The section of the basic skeleton	The information contained in it	The places in the rest of the document where the information is repeated
Executive Summary *or* Summary *or* Abstract	Undetailed overview of the *whole* document	Throughout the document
	The main conclusion(s)	Conclusions
	Possibly: the main recommendation	Recommendations
Conclusions	Overview of the conclusions you draw throughout the document	Possibly elsewhere in the document
		The main conclusion(s) will also be repeated in the Executive Summary
Recommendations	A list of your recommended actions	The main recommendation(s) might be repeated in the Abstract or Summary or Executive Summary
Appendixes		Summaries of the Appendix material might appear in the main body of the document

- For executive audiences, the diamond structure can be modified by bringing all the overview information to the front of the report.
- For nonspecialist readers, indicate to them that they can obtain undetailed overview information by reading the Executive Summary, Recommendations, and Conclusions together with the section summaries at the beginning of each section.
- Deliberate, controlled repetition of information is an essential part of the navigational pathway in an effective document.

3

Organizing a Document and Choosing Appropriate Sections

This chapter deals with planning a document and choosing a complete set of appropriate section headings. It describes the steps in organizing a document. It lists the possible sections needed for the basic skeleton of a document, together with the other sections needed to expand that skeleton into more complicated documents; it groups them into sections commonly found at the beginning and end of a document, together with others that may be needed. The lists may therefore be used to choose sections that are relevant to the particular document under construction.

Structure of the chapter

3.1   Why plan?
3.2   Steps to take when planning a document
3.3   Using the Outline mode of Microsoft Word®
3.4   Brief descriptions of possible sections to choose for a document

3.1   Why plan?

Many people feel the need to actively type or write when they start producing a document. However, the overall amount of time to final document production is greatly reduced if you spend enough time in the planning process. With an adequate structure, you will find both that it is easier to start the writing process and that the first draft will need much less revision.

3.2   Steps to take when planning a document

Step 1. Identify the main point(s) of the document from the point of view of the readers.
The aim when writing a report should be to present the information in the way the readers need it.

Three techniques to help with this are:

▶ Ask yourself "Which questions does the reader most want to have answered?" Place yourself in the readers' minds and ask yourself "Why? What? Where? When? Who? How? What next?" Depending on the topic, some of these may not be needed; some may be needed more than once.

▶ Use the technique of the Six Hidden Words: the technique taught to journalists to help them compose the first, or summary, paragraph of a newspaper article. Write down "I want to tell you that...." Then think about your readers and work out how to finish the sentence so that it identifies the aspects they will most want to know about.

> *I want to tell you that* . . . I have identified three ways in which this design should be modified.

▶ Write down the main point, keep it in front of you, and refer to it frequently throughout the writing process.

Step 2. Prepare tentative headings.
Choose headings that will signal the key areas to be covered. Use Table 3-1 to choose the various headings. Chapter 7 gives suggestions about how to structure specific types of reports.

Step 3. Decide on the subheadings to come under each heading.

Step 4. Decide on the order of your headings.

Step 5. Keep adjusting headings and subheadings.

Step 6. Collect keyword reminders of each of the points to be made under each tentative heading.
Do not collect full sentences or phrases. Do not worry about order.

Step 7. Write a mini-conclusion for each heading.
Ask "What does this mean?" The mini-conclusion is the final statement for that section. This will help to develop your logical line of argument throughout the report.

Step 8. Prepare your tentative conclusions.
At this stage, it helps the logical process by setting out the conclusions you can draw from the material in your report. However, you almost certainly will alter these conclusions during the process of writing the document.

Step 9. Prepare your tentative recommendations (if needed).
Make sure that the recommendations grow logically from the conclusions of your report. Again, you probably will revisit these.

3.3   Using the Outline mode of Microsoft Word®

It is well worth becoming familiar with the Outline mode of Microsoft Word®. It is very simple to use and is of great help in the planning and revising processes. It will enable you to:

▶ **Organize a list of headings and subheadings of various levels.**
This is useful for the first stage of organizing a document. You decide on your listing of headings, the subheadings, and their divisions. You then assign them to their various levels (level 1 for a main heading, level 2 for a subheading, etc.). They can be easily reassigned to different levels at any time in the writing process.

The text is then inserted under the headings to produce the full document.

▶ **Collapse the document to display only selected levels of headings.**
This gives an overview of the whole document. You can select the level of overview. By collapsing the document and selecting to list only the level 1 headings, you can check the overall structure of the document in terms of only its main headings. By progressively displaying greater levels of subheadings, you can obtain an increasingly more detailed view of the structure of the document.

This also helps in revising the first draft of the document (see Chapter 16).

▶ **Drag and drop a heading to a different place in the document or to a different level.**
This helps to organize and revise the document. When a heading is dragged and dropped, its associated text is also moved. This helps to avoid the errors that can arise during cutting and pasting.

▶ **Automatically produce a Table of Contents with the corresponding page numbers.**
This avoids the tedious final process in the production of a document where you have to check a document's integrity: checking that the section headings and page numbers correspond to those listed in the Table of Contents. With Word, you can produce a Table of Contents with three keystrokes.

3.4   Brief descriptions of possible sections to choose for a document

Use Table 3-1 to determine the sections to choose for a document.

The listing of the section headings in the table is divided into:
- The basic skeleton of sections
- Commonly used preliminary sections of a document
- Commonly used sections at the start of the main body of a document
- Commonly used sections at the end of a document
- Other possible sections, in alphabetical order

Table 3-1. *Possible sections to choose for a document*

Section heading	Purpose of section
The basic skeleton of sections	
Title	Gives the reader immediate access to the main subject matter.
	Running title (if needed): The short title required by journals for the tops of the pages.
Title page	Gives the title of project, project/job reference number, organization/consortium name, logos, graphics.
Executive Summary *or* Summary *or* Abstract	Identifies basic content quickly and accurately; it should be a miniaturized version of the document. It should give a brief overview of all of the key information; this is profoundly important in helping the reader assess the information in the rest of the document. It also helps readers decide whether they need to read the whole document.
	For a proposal: Gives succinct overview of whole document: all necessary information to enable the bid evaluation team to make a favorable assessment of the proposal. Should as a minimum contain: project description, approach, methodology, work program, project team, financial details, deliverables, OSH and quality management systems, conclusions (critical success factors that increase the appeal of your organization).
Introduction *or* Background	*Introduction:* Introduces the organization or consortium; location plan; explanation of the structure and contents of the document.
	Background: Gives the background to the material in the document, e.g., history of the various components of the system/equipment/area of land, etc.; present usage; concerns of the public, management, and so on.
	Either section without the other: All of the above.

Table 3-1. *Possible sections to choose for a document*

Section heading	Purpose of section
Conclusions	Lists the conclusions that you draw from the material in the report.
Recommendations	Proposes a series of recommendations for action. They should arise logically from the Conclusions.
Conclusions and Recommendations	Includes the requirements for both the above sections.
Commonly used preliminary sections of a document	
Letter of Submission *or* Letter of Transmittal *or* Cover Letter	Identifies topic and scope of the proposal document; highlights key features; overviews structure and contents; identifies the person who authorized the document; gives formal authorization of submission date. Can be included at the beginning of the document or transmitted separately.
Preparation, Review, and Authorization Page *or* Quality Record Sheet	Shows the people responsible for preparing, reviewing, and authorizing the document. Often placed immediately after the title page.
Keywords	Lists a group of words or phrases that will be used by electronic indexing and abstracting services.
Acknowledgments	Thanks the people who have given you help in your work and in the preparation of your document.
Table of Contents	Lists section headings and corresponding page numbers.
List of Illustrations *or separately as* List of Figures List of Tables	Lists—separately from the Table of Contents—the numbers, titles, and corresponding page numbers of all your figures and tables.
Glossary of Terms and Abbreviations *or* List of Symbols	Defines terms, abbreviations, acronyms, and mathematical and other symbols.
Commonly used sections at the start of the main body of a document	
Standards	Describes the standards used.
Objectives *or* Purpose Statement	Describes the aims of the project.
Scope *or* Scoping Statement	*For a feasibility study or recommendation report:* Gives the criteria you used to formulate the requirements. *For other types of reports:* Specifies the boundaries or limits of the project.
Procedure *or* Procedure Statement	Describes the processes you followed in investigating the topic of the document.
Problem *or* Problem Statement	Describes the problem and its significance.

Table 3-1. *Possible sections to choose for a document*

Section heading	Purpose of section
Commonly used sections at the end of a document	
References *or* List of References	Lists the works that you have cited in the text. For a journal or conference paper, there are strict conventions.
Bibliography	Lists the works that you have not cited in the text, which you think will be of interest to the reader.
Appendix *or* Appendixes *or* Attachments	Appears at the end of a document—for complex material that would interrupt the flow if it were inserted into the main body (for example large drawings and charts, schedules, design verifications, equipment lists, product descriptions, supporting documents).
Other possible sections, in alphabetical order	
Approach and Methodology	Describes how you intend to execute the project. This section could have three elements: List of Tasks, Technical Methods, Key Issues.
Asset Management and Gap Analysis	Shows the client that your organization has asset management information, systems, and processes in place. That will ensure the client's long-term asset performance meets market, shareholder, and legislative requirements.
Association Arrangements	Describes arrangements for association with other companies. Includes initialed copy of Terms of Reference and signed copy of other documents.
Comments on the Terms of Reference (TOR)	Summarizes your proposals with respect to the Terms of Reference; includes additional tasks, optional tasks, ambiguities or inconsistencies in the Terms of Reference, potential problems that may not have been noted in the Terms of Reference.
Company Description and Experience	Describes the organization's/consortium's general capability and experience. *If appropriate, it can be split into two separate sections:* Company Description and Company Experience
Confidentiality Clause	Ensures that information released to the client remains confidential.
Discussion	Shows the relationships among the observed facts that have been presented in the document and draws conclusions.
Facilities and Resources	Describes physical facilities such as office space and equipment, software resources, and additional staff members, organizations, and resources that could become involved in the project if necessary.
Headers and Footers	Contains enough information to remind readers of the origin of the document and what point they have reached in it.
Health and Safety (OSH) Issues	Shows the client that you have considered the various hazards that exist when carrying out this project.

Table 3-1. *Possible sections to choose for a document*

Section heading	Purpose of section
Intellectual Property *or* Client Ownership	Protects intellectual property developed by your organization in the course of the project.
Local Facilities and Counterpart Staff	Describes the counterpart office and support requirements, e.g., local staff, accommodation, equipment, and information. Who will provide them?
Materials and Methods *or* Procedure	Describes your experimental procedures. Aim: repeatability by another competent person.
Outcome	Describes the results of the event or project.
Personnel	Lists the key personnel and a summary of their qualification/ experience, roles, responsibilities, task matrix, team organization, backstopping personnel, list of technical support staff, CVs summary table. Refer reader to staff CVs later in the proposal document.
Project Understanding and Objectives	Outlines how the project fits into the client's overall program; provides more background information to the proposed approach and methodology.
	Can include master schedule/master plan.
Quality Assurance	Shows that the organization has effective quality assurance (QA) systems in place.
Records Review	Describes the records relevant to the project.
Reference Documents	Identifies all documents that relate to the study, such as drawings, standards, specifications, and handbooks.
Response Matrix	Itemizes the features of the Terms of Reference, and refers to the page number in the proposal where each is addressed.
Results	Presents the results of an investigation.
Risk Assessment *or* Risk Management	Lists the perceived risks and describes the methodology used to assess risk; assessments of the risks; risk minimization or mitigation.
Site Appreciation	Describes the site, provides technical notes, photographs.
Staff CVs	Provides current CVs for each staff member involved in the project.
Standards Used	Describes standards and codes of practice that will be adhered to.
Statutory Framework	Describes the regulations relevant to the project.
Technical and Management Skills	Describes technical skills, experience, availability, and commitment of team members.
	Describes management skills of team members, including client liaison, reporting structure, previous performance, qualifications, training and practical experience, proposed management systems.

Table 3-1. *Possible sections to choose for a document*

Section heading	Purpose of section
Time-Based Diagrams *or* **Charts**	Presents charts to describe the interrelationship of tasks required in the work program and shows how this can be achieved in the timeframe.
Work Program, Staffing Schedule, and Organization *or* **Schedule**	Describes the various schedules: organization chart, staffing schedules, critical path chart, man-month estimates, work program, key personnel tasks, time-based diagrams.

4

Presentation Style

.

Readers are only human; if any document makes a good first impression and if the information can be easily found (see Chapter 2), then it tends to be more highly rated overall. This holds true even though official evaluation systems of proposals tend to award only about 2% of the total points to the quality of presentation.

An organization's presentation style is usually governed by its house style. Therefore only general indications of page layout, orientation, font, and binding can be given here.

Page layout

Pages of a document should appear as though they have not been crammed with information. The elements of this are a large binding left-hand margin, at least 1-inch margins top and bottom, and intelligent use of white space. However, avoid unnecessary large areas of white space: The reader may think you are trying to artificially expand the size of the document.

Headers and footers should contain enough information to remind readers of the origin of the document and what point they have reached in it. See Section 5.6.10.

Page orientation

Pages can be printed either with portrait orientation (long axis vertical) or landscape orientation (long axis horizontal). Portrait orientation is usually preferred, although tables and graphics may sometimes need to be printed landscape. In this case, make sure that the reader does not constantly need to turn the document through 90°. The convention states that any such page should be read from the right-hand (outer) edge of the page.

Font

Font type

Many companies now have a house style to maintain a corporate image. If your company does not have one, choose either a plain serif font (e.g., Times Roman, Palatino) or a sans serif font (e.g., Arial, Trebuchet). A serif font is believed to be more appropriate for rapid reading; however, many companies choose a sans serif font because of its clean lines. Avoid elaborate fonts and the old-fashioned typewriter-face Courier.

The general rule is to use no more than two fonts in a document. An elegant method is to use, for example, a serif font for the text and a sans serif font for headings.

Use the same font type for both the main document(s) and the cover letter.

Type size

The general rule for the main text is to use 11- or 12-point type. Less than 10 point can be used in tables (see Chapter 16), but it is too small for large blocks of text. Headings can be of slightly larger font, but this should be done discreetly.

Justified text

Many organizations choose to use justified text—text that is aligned at both the left-hand and right-hand margins—because it looks more organized. However, text that is aligned to only the left-hand margin—technically known as ragged-right—is easier to read, because there is equal spacing between the words.

Page numbering

See Chapter 15.

Binding

Lever-arch binders may be more convenient for large documents produced by a team, where changes are likely to be made up to the final moments before submission. For the convenience of the reader, use divider sheets that are labeled to correspond to the section numbers.

The Sections of a Document

Part 2 consists of one large chapter that explains the requirements for sections and elements of a document. Most of the sections likely to be found in a representative variety of engineering documents are included. Most are described in terms of the purpose of the section and how to write it, with relevant examples and other notes where necessary.

5

Requirements for Sections and Elements of a Document

This chapter covers the requirements for each of the sections and elements found in representative types of engineering documents. Most sections are described in terms of the purpose of the section and how to write it, with relevant examples and other notes where necessary.

Structure of the chapter

5.1   Listing of commonly used sections and elements of a document
5.2   Requirements for the basic skeleton of sections
5.3   Requirements for commonly used preliminary sections
5.4   Requirements for sections commonly used at the start of the main body of the document
5.5   Requirements for sections commonly used at the end of a document
5.6   Requirements for other possible sections, in alphabetical order

Notes   For a specific type of document, use the relevant chapter for that document type in combination with this chapter.

Any one document will not need all of the sections described in this chapter.

It is essential to check whether your organization has specific requirements for the sections of the document.

5.1 Listing of commonly used sections and elements of a document

The following sections and elements of a document are covered in this chapter.

Basic skeleton of sections
Title
Title Page
Executive Summary *or* Summary *or* Abstract
Introduction *or* Background *or* both
Conclusions
Recommendations
or Conclusions and Recommendations

Preliminary sections
Letter of transmittal *or* Cover letter
Preparation, review, and authorization page *or* Quality record sheet
Table of Contents *or* Contents page
List of Illustrations
Keywords
Acknowledgments
Glossary of Terms and Abbreviations *or* List of Symbols

Sections at the start of the main body of the document
Standards
Objectives *or* Purpose *or* Purpose Statement
Scope *or* Scoping Statement
Procedure *or* Procedure Statement
Problem *or* Problem Statement

Sections at the end of a document
Bibliography *or* References *or* List of References
Appendix *or* Appendixes *or* Attachments

Other possible sections and elements, in alphabetical order
Approach and Methodology
Asset Management and Gap Analysis
Association Arrangements
Comments on the Terms of Reference (TOR)
Company Description and Experience
Confidentiality Clause
Discussion
Facilities and Resources
Headers and Footers
Health and Safety (OSH) Issues
Intellectual Property *or* Client Ownership
Local Facilities and Counterpart Staff
Materials and Methods *or* Procedure

Outcome
Personnel
Project Understanding and Objectives
Quality Assurance
Records Review
Reference Documents
Response Matrix
Results
Risk Assessment *or* Risk Management
Site Appreciation
Staff CVs
Standards Used
Statutory Framework
Technical and Management Skills
Time-Based Diagrams or Charts
Work Program, Staffing Schedule and Organization *or* Schedule

5.2   Requirements for the basic skeleton of sections

5.2.1   Title

Purpose
To adequately describe the contents of your document in the fewest possible words.

Guidelines for writing it
For the rules of capitalization in a title, see Section 15.2.
- ▶ Identify the information that a reader would need to gain immediate access to the main point of your document.
- ▶ Ensure that the title is not too general or too detailed, and that it contains the necessary key information.
- ▶ Ensure that it makes sense. The structure can be lost during the quest for the minimum number of words.
- ▶ If the title is a long string of words, rewrite it into two parts using a colon (see Section 12.3.1).

Running head
The short title required by journals for the tops of the pages (see Chapter 12).

5.2.2   Title Page

This is also called the edge page, if using a lever-arch binder.

Purpose
To attractively present the title of project, project/job reference number, organization/consortium name, logos, and graphics.

Guidelines for writing it
Items of information that you may wish to include are:
- The project title (preferably the same as that used in the TOR)
- The company logo
- The client logo
- The project reference number (usually given in the TOR)
- The name of the company or consortium
- A key color photograph or graphic

5.2.3   Executive Summary *or* Summary *or* Abstract

Summaries are critical for the understanding of the whole document. They are described in detail in Chapter 6.

5.2.4   Introduction *or* Background *or both*

Important: Do not include an Executive Summary in the Introduction, as some authorities advise. It defeats the purpose of a Summary (see Chapter 2).

There is always interplay between the Background and the Introduction. When free to choose, some companies prefer one or the other, whereas others include both.

To write either a Background or an Introduction
Combine the guidelines given here for both the Introduction and Background sections. Alternatively, the Introduction could contain the subheadings *Background* and *Scope* (see Section 5.4.3).

To include both Background and Introduction
The differences in general terms are:
- The Background gives the history of the subject matter and the objectives of the study. Alternatively, the objectives can be stated in a separate Objectives section.
- The Introduction usually restates the brief and describes the structure of the report.

Introduction

Purpose
To give an introduction to the organization or consortium, explanation of the structure and contents of the document, and a listing of other relevant documents.

Guidelines for writing it
- ▶ Describe the history of events leading to the proposal.
- ▶ List the member companies and explain what each brings to the project.
- ▶ Identify the lead company.
- ▶ List the various documents that form part of the TOR, and any later amendments.

- ► List other documents relevant to the project (e.g., environmental impact assessments, standards, etc.).
- ► Briefly describe the structure of the proposal—the headings and contents of each of the main chapters. Think of it as a route map; show the clients where they are going to be led and the overall structure of the document. Explain briefly what is going to be explained where.

Background

Purpose
To give the background to the material in the document (e.g., history of the various components of the system/equipment/area of land, etc.); present usage; describe concerns of the public, management, and so on.

Guidelines for writing it
Make sure that if you are writing only a Background section (i.e., there is no Introduction), then the Background contains the combined information for both an Introduction and a Background.

Describe the background to the documented work, covering such issues as history of the various components of the system/equipment/area of land, and so on; present usage; describe concerns of the public, management, and so on; and explain other relevant issues.

Some TORs call for a section called Background and Organization. In this case, in addition to the material included in a Background section, also provide an introduction to the background of the organization or consortium and describe its general capability and experience.

5.2.5　Conclusions

Purpose
To present conclusions that arise from the material in the document. Each conclusion *must* be substantiated.

Note: In a professional document, the Conclusions section is regarded as another aspect of the summarizing process. The rule of thumb should be that a person needing a complete, undetailed overview of the document should be able to read the Title, Summary *or* Abstract, Conclusions and Recommendations (see Chapter 2).

Guidelines for writing it
Important: There should be no new material in this section. Each conclusion must be drawn directly from material that has already been presented in the main body of the report and needs to be well substantiated.
- Each conclusion should be related to specific material.
- Each conclusion should be brief (since the full explanation is given elsewhere in the document).
- A numbered or bulleted list can be used if appropriate. Start with the main conclusion, and then present remaining conclusions in descending order.

- If there are many conclusions, they can be grouped under headings as shown below. In this case, ensure that the numbers run sequentially through the whole list. This makes them more readily identifiable.

Example

Conclusions

Public health issues

1. The waters of the basin continue to present a significant risk to human health owing to sewage pollution.
2. Sewer overflows are a major source of pollution; they contribute over 98% of the fecal coliform pollution load to the basin, and half the nitrogen load.
3. *[etc.]*

Water quality

4. The water in the basin is very highly loaded with nutrients, mainly from sewage overflows.
5. Stormwater is also a major source of pollution....
6. *[etc.]*

Siltation

7. Sedimentation plates indicate that the level of siltation in the basin has increased from 3 to 6.5 mm per year over the past 20 years.
8. Sediment quality is fair and is not considered to present an abnormal health risk....
9. *[etc.]*

5.2.6   Recommendations

Purpose

To propose a series of recommendations for action as a result of the conclusions drawn from your work.

Position in the document

In a formal engineering document, placement of the Recommendations section is usually one of the following:

- At the end of the document, after the Conclusions. In this case, it can often be usefully combined with the Conclusions into a section called Conclusions and Recommendations.
- At the start of the document, immediately after the Summary (see Chapter 3).

Guidelines for writing it

- Recommendations are your subjective opinions about the required course of action. Make sure, though, that the wording is objective and not opinionated.
- Recommendations are often best given as a numbered list. Each item should be brief.
- Make the main solution to the problem your first recommendation. This usually fulfills the purpose of the report. Then list the other recommendations in a logical way.
- No recommendation should be unsupported. The supporting information should exist elsewhere in the document.

- Avoid the legalistic form that strings together a number of recommendations.

Example

Recommendation

That proposed Scheme Change 5 to the Operative Newtown District Scheme, Proposed Change 44 to the Operative Middletown District Scheme, and Proposed Change 2 to the Third Review of the Uppertown District Scheme (when made operative), an example of which appears as an Appendix to this report, be publicly notified.

Rewritten

Recommendation

That the following scheme changes be publicly notified:
1. Proposed Change 5 to the Newtown District Scheme
2. Proposed Change 44 to the Operative Middletown District Scheme
3. Proposed Change 2 to the Third Review of the Uppertown District Scheme (when made operative)

Tense of the verb

► Write Recommendations in the conditional, subjunctive, or present form, or as a series of instructions in the imperative form (see Section 17.9.4).

Examples

It is recommended that:
The operation of the sluice gates should be automated *(conditional)*.
The operation of the sluice gates be automated *(subjunctive)*.
The operation of the sluice gates is automated *(present)*.

or

The recommendations are:
Automate the operation of the sluice gates *(imperative)*.

► Ensure that only one form of the verb is used in a set of recommendations.

Example: Poorly written Recommendations section made up of mixed forms of verbs

1. Develop a management plan for the West Basin in consultation with residents.
2. Formation of a community committee made up of residents and government staff.
3. Regular street cleaning.
4. Inspection, maintenance, and repair of screens on sewer overflow manholes.
5. Maintain sluice gates to reduce leakage.

Rewritten as a series of instructions (imperative form of the verb)

1. Develop a management plan for the West Basin in consultation with residents.
2. Form a community committee made up of residents and government staff.
3. Clean the streets regularly.
4. Inspect, maintain, and repair screens on sewer overflow manholes.
5. Maintain sluice gates to reduce leakage.

Rewritten using the conditional tense of the verb

1. A management plan for the West Basin should be developed in consultation with residents.
2. A community committee made up of residents and government staff should be formed.
3. The streets should be cleaned regularly.
4. The screens on sewer overflow manholes should be inspected, maintained, and repaired.
5. The sluice gates should be maintained to reduce leakage.

5.2.7   Conclusions and Recommendations

The material for both a Conclusions section and Recommendations section can be usefully combined into this one section. It has the advantage that the individual recommendations are related more immediately to the relevant conclusions.

5.3   Requirements for commonly used preliminary sections

5.3.1   Letter of Transmittal *or* Cover letter

Purpose

Identifies topic and scope of the proposal document, highlights key features, overviews structure and contents, identifies the person who authorized the document, and gives formal authorization of submission date (see Chapter 9).

5.3.2   Preparation, review, and authorization page *or* Quality record sheet

Purpose

To show the people responsible for preparing, reviewing, and authorizing the document. Usually a simple three-line list:

- Prepared by . . . *(signature and name)*
- Reviewed by . . . *(signature and name)*
- Authorized by . . . *(signature and name)*

It is often placed immediately after the title page.

5.3.3   Table of Contents *or* Contents Page

Purpose

To list the section headings and subheadings, together with their corresponding page numbers.

The problems of layout, formatting, and correspondence of page and section numbers can be eliminated by using the facility on your word processor that automatically constructs and formats your Table of Contents (see Section 3.3).

Guidelines for writing it

If you are not using the word processor's facility, use the following guidelines:

- ▶ Decide the lowest level of heading to display in the Table of Contents (e.g., whether you want to go down to subheading level or to subsubheading level).
- ▶ List all the sections and all their subheadings to your chosen level down the left-hand side of the page.
  - Number the sections and their subheadings by the accepted conventions, using the decimal point numbering system (see Chapter 15).
  - Any indentations for subheadings should be consistent for each level of heading.
- ▶ Place the corresponding page numbers at the right-hand side of the page. For the conventions for numbering the pages, see Chapter 15.
- ▶ Individual illustrations are not listed in a Table of Contents. If the document has many illustrations, include a List of Illustrations (see below) immediately after the Table of Contents.
- ▶ Conventionally, the Abstract or Summary is not listed in the Table of Contents. However, it may help the reader to do so, even though it is placed immediately after the Title Page and is therefore easily found.

5.3.4   List of Illustrations

Purpose

The term *illustrations* includes tables and figures (graphs, line drawings, photographs, maps, etc.). To give a listing—separate from the Table of Contents—of the figure and table numbers, their titles, and corresponding page numbers. For the conventions for figure and table numbering, titling, and captioning, see Chapter 15.

Guidelines for writing it

- ▶ Use the title List of Illustrations if your document contains both tables and figures. If it contains only tables, call it List of Tables; if only figures, List of Figures. (This book has so few illustrations that a List of Illustrations is unnecessary.)
- ▶ If you are using a List of Illustrations, list all the figures first and then list all the tables.
- ▶ List the number, title, and page of each illustration.
- ▶ Place the List of Illustrations immediately after the Table of Contents. If both of them are brief, put them on the same page, with the Table of Contents first.

5.3.5   Keywords

A list of keywords used in electronic searching (see Section 12.3.4).

5.3.6   Acknowledgments

A courteous acknowledgment of contributions from people or organizations.

5.3.7   Glossary of Terms and Abbreviations *or* List of Symbols

Many reports assume that the readers will be familiar with acronyms, abbreviations, and other terms that are second nature to the writer. The reading process is greatly helped if a glossary of such terms is included.

Purpose

To define the specialized terms, symbols, and abbreviations (including acronyms) used in the main text of the document.

Guidelines for writing it

Terms that need to be dealt with include:

- Specific technical terms.
- Greek or other symbols.
- Abbreviations that are not commonly known.
- Acronyms. These are often in the form of the initial letters in capitals of a series of words, e.g., ESI: environmental sustainability index; FEA: finite element analysis. For the conventions for also defining these in the text, see Section 15.3.

Before you list the terms and abbreviations, it may be appropriate to state:

> S.I. (Système International d'unités) abbreviations for units are used in this work. Other abbreviations are listed below.

Where to place it

The Glossary of Terms can be placed either at the beginning of the document immediately after the Table of Contents or the List of Illustrations (this is the optimal position for the reader). Alternatively, a fold-out sheet can be placed at the back of the document so that it is readily available during the whole reading process.

If the glossary is large and you think that it needs to be at the end of the document, readers would appreciate a note placed immediately before the Introduction, referring them to the page number of the glossary. Suggested wording:

> Explanations of terms and abbreviations used in this document are given in the Glossary of Terms and Abbreviations, page #.

5.4   Requirements for sections commonly used at the start of the main body of the document

5.4.1   Standards

Purpose

To describe the standards used.

Guidelines for writing it

List the standards, and describe the reasons for using those particular ones. State any reasons for deviating from the standards requested by the client.

5.4.2   Objectives *or* Purpose *or* Purpose Statement

Purpose

To describe the aims of the project.

Guidelines for writing it

▶ State the objectives clearly, briefly, and concisely. It is often useful to state the primary objective and then follow with the specific objectives. For example:

The past and present operations at the site and its surroundings were reviewed with the primary objective of assessing key environmental liabilities. Specific objectives were as follows: *(followed by the listed specific objectives)*.

In a longer document, it is effective to state each broad objective and then list under each one its relevant specific objectives.

▶ Name the alternatives if necessary.

5.4.3   Scope *or* Scoping Statement

Purpose

- *For a feasibility study or recommendation report:* To name the criteria you used to formulate the requirements.
- *For other types of reports:* To specify the boundaries or limits of the project.

Example

1.2 Project Scope

(Name of organization) was engaged by *(name of organization or person)* to carry out the following investigations:

1. To review the existing investigation into the disposal of stormwater from a landlocked area through a proposed stormwater retention pond located in Middle Park.
2. To investigate interaction with existing reticulation immediately downstream of Middle Park.

3. To carry out preliminary investigations at the proposed site for the retention pond in Middle Park.
4. To assess the feasibility of an alternative site for the proposed retention pond.

5.4.4   Procedure *or* Procedure Statement

Purpose
To describe the processes you followed in investigating the subject of the report.

Guidelines for writing it
> ▶ *For an experimental investigation:* See Section 12.3.6.
> ▶ *For other types of reports:* Explain all the actions you took: the people you interviewed, sites investigated, research performed, and so on.

5.4.5   Problem *or* Problem Statement

Purpose
To describe the problem and its significance.

Guidelines for writing it
Probably in this order:
1. Describe the problem, giving the basic facts about it.
2. Explain what has gone wrong.
3. Specify the causes or the origin of the problem.
4. Describe the significance of the problem (short-term and long-term).
5. Give the appropriate data and state their sources.
6. Specify who is involved and in what capacity.
7. Discuss who initiated action on the problem or what caused you to write the report.

5.5   Requirements for sections commonly used at the end of a document

5.5.1   Bibliography *or* References *or* List of References

See Chapter 14.

5.5.2   Appendix *or* Appendices *or* Attachments

Purpose
To provide, at the end of the document, complex material that would interrupt the flow of the document if it were to be inserted into the main body (large drawings and charts, schedules, design verifications, equipment lists, product descriptions, supporting documents, and so on).

How to assemble the Appendixes
- Material included in an appendix should be there for a specific purpose.
- They should contain well-structured information.
- Related material should be grouped into separate Appendixes.
- Do not include company brochures.
- Give each Appendix a number or a letter, followed by a title (e.g., Appendix 3: Equipment lists). For the conventions for numbering Appendixes and their page numbers, see Section 15.4.
- The number and title of each appendix should be listed in the Table of Contents (see Section 5.3.3).
- Every item that is included in an appendix should be referred to at an appropriate place in the text.

5.6   Requirements for other possible sections, in alphabetical order

5.6.1   Approach and Methodology

Note: See also Sections 5.4.4 and 12.3.6.

Purpose
To describe how you intend to execute the project.

Guidelines for writing it
This section is commonly found in a proposal. It could contain subsections. Topics to be covered could include:
- The various options
- Your proposed approach
- Ranking of alternatives
- Issues
- Answering client needs
- The list of deliverables
- Liaison points where you will refer to the client
- Tasks that are additional to those stated in the TOR. Cross-refer to the section, Comments on the Terms of Reference.
- Summary flowcharts and tables.

It may be appropriate to structure it in the following subsections:
- *Task list*: Lists of the various tasks.
- *Technical methods*: In-depth comment on options, proposed approach, and methodology; ranking of alternative concepts; optional services; project management methods; list of deliverables.
- *Key issues:* Environmental monitoring; training/technology transfer; computer hardware/software; project management; backstopping methods; support for minorities; quality assurance; health and safety.

- *Work Program, Staffing Schedule, and Organization Chart* can also form a subsection or can stand alone as a separate section.

Task List

Purpose
- To show the client how you propose to carry out the tasks specified in the TOR and others that you believe are necessary.
- To show how the tasks will be distributed between the companies and individual personnel.
- To create the structure around which to build the rest of the document.

Guidelines for writing it
- Before writing, clearly define and prioritize the tasks to avoid overwhelming the client.
- Make sure that the tasks are as specified in the TOR. If you are defining additional tasks that are not specified in the TOR, show whether you consider each one to be essential or optional to the project. Cross-refer to Comments on the Terms of Reference.
- Link the task list to a summary table and/or flowchart.
- In the task list table/flowchart:
  — Describe each task briefly (10–15 words). Use the same words as those used in the Terms of Reference. This will help the reviewer to compare your proposal with the tasks specified in the TOR.
  — Structure the list of tasks either in terms of main tasks and subtasks or in terms of project phases.
- If necessary, show which company is responsible for which task.

Technical methods

Purpose
- To discuss the technical tasks of the project.
- To describe the technical methods you will use to carry out the tasks in the task list and to explain why you have chosen these methods. You may also need to include reference to the standards and codes of practice used (see Section 5.6.25)

Guidelines for writing it
Important: If you are working in a consortium, this section may involve the collation of material from the various companies. Substantial rewriting may therefore be necessary to achieve a consistent style.

The task list has already defined what to write about. The technical and (if an international proposal) country experts will now decide what to write.

The notes should contain the following elements:
- A description of each basic task.
- A description of the approach you propose using.

- A description of local issues such as (a) local knowledge (project history, client organization, local authorities involved in approval of the design), and (b) local content (national design standards, local conditions bearing on the design). This section can be linked back to the Site Appreciation and Project Appreciation sections. Avoid using modules for this section; good technical notes should show that you are defining the project in terms of the local conditions.

Key issues

Purpose
To highlight the aspects of your project that you believe are particularly important to you and to the client.

Guidelines for writing it
Important: You will need to define the themes that are important to the client or funding agency. They can be identified by being specifically mentioned in the TOR or in the client's or funding agency's publications, or they can be emphasized by the client in other ways.

Possible key themes
Many of these appear frequently; it is worth preparing standard modules that can be modified to fit each individual project. Frequent themes are:
- Backstopping
- Computer hardware/software (alternatively, this can be included in Facilities and Resources)
- Consideration of vulnerable groups (e.g., rights of and interaction with indigenous peoples, support for minorities, resettlement monitoring). Funding agents such as the World Bank regard consideration of vulnerable groups as a significant factor in the quality of a proposal.
- Cost-effectiveness
- Environmental protection/monitoring
- Health and safety (see Section 5.6.10)
- Project management
- Quality assurance (see Section 5.6.17)
- Stakeholder involvement
- Technology transfer.

Work Program, Staffing Schedule, and Organization Chart

See Section 5.6.29.

5.6.2   Asset management and gap analysis

Purpose
To show the client that your organization has asset management information, systems, and processes in place that will ensure that the client's long-term asset performance meets market, shareholder, and legislative requirements.

Guidelines for writing it

- ▶ Give an assessment of the existing records and asset information available to support ongoing operations, maintenance, and renewal activities.
- ▶ Describe the information systems that your organization uses to store, analyze, and report asset information.
- ▶ Give an assessment of the practices and processes used to manage assets.
- ▶ Provide operation and maintenance analysis and strategies.
- ▶ Provide risk management strategies.
- ▶ Provide asset renewal forecasts and strategies.
- ▶ Describe asset optimization strategies.
- ▶ Describe decision-making processes.
- ▶ Describe your organization's key steps in this process, such as:
  - Reviewing relevant documentation such as asset management plans, maintenance plans, and decision-making processes.
  - Interviewing relevant staff of the client organization.
  - Evaluating the client organization's information, systems, and processes against appropriate industry best practice.
  - Including a gap analysis chart listing the status of activity against appropriate best practice.
  - Identifying shortfalls in practice and performance, and the relevant improvement procedures required to achieve industry best practice.

5.6.3   Association arrangements

Purpose

To describe the proposed association between the companies within a consortium and the division of the work.

Guidelines for writing it

- ▶ Describe the arrangements between your organization and the others in the consortium. Clearly show the relationships between the various organizations—whether you have a joint venture, subconsultancy, or freelance arrangement.
- ▶ The Terms of Reference usually require written evidence that the lead consultant has a documented agreement with the other companies. However, you may also need to discuss with the client whether more material is needed, for instance, an explanation of how the work will be divided.
- ▶ Include an initialed copy of the Terms of Reference and signed copies of other documents.

5.6.4   Comments on the Terms of Reference (TOR)

Purpose

- To summarize, but not discuss, your proposals with respect to the TOR.
- To record the additional tasks that you believe are necessary and any optional tasks.

- To point out ambiguities or inconsistencies in the TOR.
- To record potential problems that may not have been noted in the TOR.

Guidelines for writing it

▶ Keep in mind that you may be pointing out the inadequacies of the client's TOR as perceived by your organization. To avoid irritating the client, this section must be written carefully, avoiding argument or false flattery. This can be helped by including a section called *Enhancements to the TOR* or including a positive statement such as:

> Notwithstanding the comments, we are satisfied that the Terms of Reference sufficiently define the scope of work and that matters relating to the execution of the work can be resolved between us during negotiations.

▶ Use this section to summarize your proposals for additional work to that cited in the TOR, for example, additional experts, additional progress reporting, additional quality assurance procedures, and so on. Discussion should be placed elsewhere in your document.

▶ For each of your proposals, indicate which part of the TOR it refers to and the place in your document where it is considered in detail.

▶ Also point out ambiguities or inconsistencies in the TOR, together with potential problems that may have been overlooked and your proposals to deal with them.

5.6.5 Company description and experience

Purpose

To show the client your organization's or consortium's total capability of carrying out the proposed work in terms of the following:

- Its established reputation.
- Its appropriate experience.
- Its appropriate experience in the client's country *(if an international proposal)*.
- Its facilities and resources.

Guidelines for writing it

If needed, this can be split into two separate sections: *Company Description* and *Company Experience*. It can also include *Facilities and Resources* (see Section 5.6.8).

Specifically answer what the TOR calls for. If you decide that more detail is necessary, make sure that it is relevant and concise.

Company description

- Business structure: to show the relationships between the various organizations in the consortium.
- Current activities: to show your activity in the relevant types of projects.
- Location of offices, particularly if the organization has offices in the project region.

- Key staff: qualification/experience of key personnel, roles, responsibilities, task matrix, team organization, management/reporting structure.
- Key issues: the issues that you have identified as being the important selling points of your proposal.

Company experience **or** *Relevant experience* **or** *Track record*

This section should show depth and breadth of experience by outlining projects of a similar nature completed by your organization. Use project references to illustrate the company's or consortium's experience of working in (a) the technical sectors and (b) the project region (if an international proposal).

- ▶ Select the project references according to what is specifically called for in the Terms of Reference, the key issues, and the direct relevance of the project.
- ▶ List the more relevant projects first, then other projects in order of date, starting in the present and working backward.
- ▶ For each project reference, give the following information: description; value; duration; client; whether they were completed on time, to budget, to a specified quality; value engineering; partnering; and contacts.
- ▶ Provide at least four contact people for each project. This increases the chance of the client's being able to readily contact at least three of them.
- ▶ Make sure that tables or graphics do not heavily break the flow of the story you are telling. They may be better placed at the end of the chapter.

5.6.6   Confidentiality clause

Purpose

To ensure that information released to the client remains confidential.

Guidelines for writing it

Important: To ensure adequate protection, legal personnel are best consulted for the wording of any such agreements.

Possible wording
Confidentiality
All information supplied to the client by *(name of organization)* is confidential between *(name of organization)* and the client. If the client wishes to release any of the tendered information to a third party (e.g., for evaluation of the tender), it can only do so with the prior written approval of *(name of organization)*.

5.6.7   Discussion

This section is almost always required in a journal paper and sometimes is appropriate in other types of documents. In a Discussion, show the relationships among your observations and place them in the context of other documentation and other people's observations and work. See Section 12.3.8.

5.6.8   Facilities and resources

This may also be incorporated into Company description and experience (Section 5.6.5).

Purpose
To describe the physical facilities, organizations, and resources that could become involved in the project if necessary.

Guidelines for writing it
The following may need to be described:
- Physical facilities: office space, equipment, computer hardware, laboratory resources, or other. Simple listing; can be supported by drawings or photographs.
- Software resources: computer system software such as modeling packages, statistical packages, library systems, and so on.
- Additional staff: other staff members who could become involved in the project if necessary. Also, other organizations and resources that can be called upon to provide help if the program requires it (e.g., details of other organizations that have agreed to back you up, available laboratory facilities, and CVs of additional staff).

5.6.9   Headers and footers

Purpose
To contain enough information to remind readers of the origin of the document and what point they have reached in it, for example:
- Company name and logo
- Project/job number and title
- Section/chapter title and number
- Page number (see Section 15.4).

5.6.10   Health and Safety (OSH) issues

Purpose
- To show the client that you have considered the various hazards that exist when carrying out this project.
- To show that these hazards have been addressed through elimination, isolation, or minimization.

Guidelines for writing it
- ▶ Give details of your current Health and Safety Management systems.
- ▶ Give details of current independent audit procedures and any accreditation that the system has.

- ▶ Give a site-specific Health and Safety plan that identifies the specific key hazards.
- ▶ Include your company's safety record for the past x years. This can include, for example:
  - Number of fatalities
  - Number of injuries resulting in one or more days off work
  - Number of near-miss accidents
  - Number of accidents resulting in environmental damage or pollution
  - Incidents of cautioning or prosecution by an enforcement authority.

5.6.11   Intellectual property *or* client ownership

Purpose
To protect intellectual property developed by your organization in the course of the project.

Guidelines for writing it
To ensure adequate protection, legal personnel are best consulted for the wording of any such agreements.

Possible wording to define intellectual property
Intellectual property shall include, but is not restricted to:
- Any proprietary software that has been developed by *(name of organization)* for their benefit.
- Any proprietary procedures, technical knowledge, or quality management systems that *(name of organization)* proposes to use on the project.

Possible wording to protect intellectual property
Any intellectual property developed by *(name of organization)* solely for this project shall pass to the ownership of the client upon full and final payment of all accounts due.

Any intellectual property that *(name of organization)* created prior to this tender remains the sole property of the party. This information or property will not be signed over for the use of the client unless one of the following conditions applies:
- The party has agreed in writing that the client will be given the property.
- It has become publicly known through no wrongful act of the client.
- It was received from a third party prior to the tender closing date.
- It has been disclosed as a requirement of law or government.

5.6.12   Local counterpart staff and facilities

Purpose
To describe and confirm the facilities to be provided by the client: local staff, local accommodation, and equipment, as well as information required from the client.

Guidelines for writing it
- *Local staff.* This could include:

— *Professional staff* such as engineers, geologists, environmental experts, sociologists, legal advisors, merchant bank representatives, accounting professionals, and others. These people should be named and their CVs should be included.

— *Technical support staff* such as drafting personnel, field assistants, laboratory staff, and interpreters.

— *Nontechnical support staff* such as administrative staff, clerical staff, drivers, and guards.

- *Local accommodation.* This could include both office and residential.
- *Local equipment.* This could include means of transport, office and residential furnishing, office equipment, computer hardware and software, and power and telephone costs.
- *Information* needed from the client at the time of mobilization.

5.6.13   Materials and methods *or* Procedure

This section is most commonly found in a paper reporting on research results. See Chapter 12.

5.6.14   Outcome

Purpose
To describe the results of the event or project.

Guidelines for writing it
- Describe the different aspects of the results of the event (e.g., as in an investigation report) or the project (e.g., as in a progress report).
- For an investigation such as a laboratory report, a Results section may be more appropriately substituted (see Chapter 12).
- For other types of reports, it may be more appropriate to instead use Conclusions and Recommendations sections.
- The action that needs to be taken can also be included in this section, but it is more commonly placed in the Recommendations.

5.6.15   Personnel

Purpose
To show the client that your selected personnel have the required qualifications, experience, and skills for the project.

Guidelines for writing it
- *Selection criteria.* Outline the criteria used to select the project team and other key personnel.
- *Proposed team.* Give details of each member of the team: summarize their qualifications and experience, roles, and responsibilities; describe technical support staff; and describe backstopping personnel. According to your decision, you can include either of the following:

— The detailed staff CVs. However, if the full CVs are included at this point in the proposal, the flow of the story may be interrupted.

— A separate section called *Staff CVs* later in the document. In this case, include a CVs summary table at this point (see Section 5.6.24).

- *Relevant experience.* Outline the individual and group experience that is directly relevant to the project.
- *Matrix of duties.* Outline the duties of each team member and show the team organization and interactions.

5.6.16   Project understanding and objectives

Purpose

- To describe how this project fits into the client's overall program.
- To outline the project's aims and objectives.
- To identify any external constraints.
- To give background information to the proposed approach and methodology.

Guidelines for writing it

You need to understand the client's overall aims and objectives. Material for this section will come from site visits and discussion with your client.

▶ Outline how the project fits into the client's overall program.

▶ Give more background information to the proposed approach and methodology.

▶ Identify the constraints on the project from external sources.

▶ Include a master schedule or plan if appropriate.

5.6.17   Quality assurance

Purpose

- To show the client the effectiveness of the organization's quality assurance (QA) systems.
- To assure the client that the product is independently formally reviewed and extensively audited throughout the project.
- To show that appropriate systems are in place to prevent any errors from continuing through into subsequent sections of the work.

Guidelines for writing it

This can be a difficult section to write in that the quality attributes can be made to look too rosy.

▶ Give a brief description of the ISO (International Standards Organization) accreditation systems that are in place.

▶ If there is a regular audit of the organization by an independent accreditation organization, give a brief description of the procedures and the credentials of the accreditation organization.

▶ Include details of how and when checks are to be made on the work and how this auditing process will be documented.

▶ *(For projects involving computer operations such as modeling)*: Describe your organization's file management systems. This is important; many computer operations contain a very large amount of input data and the resulting output files can stretch the capacity of many computer systems.

▶ State that your organization has rigorous QA procedures in place and that a summary of procedures and QA certificates are available on request.

5.6.18   Records review

Purpose
To describe the records relevant to the project, such as publicly available databases and documentation.

5.6.19   Reference documents

Purpose
To identify all documents that relate to the study, such as drawings, standards, specifications, and handbooks.

5.6.20   Response matrix

Purpose
To provide an aid that helps the client to readily assess whether the proposal addresses each item in the Terms of Reference and to navigate through a document that can be large and complex.

Note   Clients may find the evaluation of different proposals difficult because of the various formats that may be used. A response matrix has been shown to be a very effective means to show the client that you have responded to all their requirements. It is placed as part of the frontmatter of the document.

Guidelines for writing it
A three-column format can be used: the first column for itemizing the elements of the TOR; the second for showing the page number(s) in the proposal where each item is dealt with; and the third left blank for the client's comments and checking off.

Suggested format

Terms of Reference	Proposal response	Client comments
p. 4: Relevant experience	pp. 13–16	
pp. 5–6: Approach and methodology	pp. 18–24	
p. 15: Estimates of local facilities	pp. 30–32	

5.6.21   Results

This section is most commonly found in a journal or conference paper or a report of a straightforward investigation (see Section 12.3.7).

5.6.22   Risk assessment *or* Risk management

Purpose
To describe the perceived risks.

Guidelines for writing it
The following elements may be needed:
- Risk assessment methodology
- Methods such as the software used, standards, and so on
- Hazard identification
- Exposure assessment: descriptions of human and environmental receptors
- Human health risk assessment and environmental risk assessment: descriptions of, for example, soil, groundwater, and vapor contamination
- Risk minimization and mitigation

5.6.23   Site appreciation

Purpose
To describe the physical nature of the project site. This provides the background for the proposed approach and methodology.

Guidelines for writing it
This is a high-priority section; the proposal team's personnel will not be able to propose suitable techniques until they know what the site looks like. Much of this information will come from the site visit and discussion with the client.

This section should consist of technical notes and photographs describing the area's project-related characteristics.

5.6.24   Staff CVs

Purpose
To give the relevant details of the key personnel to be involved in the project.

Guidelines for writing it
- Even though staff CVs are usually kept on file, make sure that the information in each one is custom-chosen so that it is relevant to the project and to what the client is looking for.
- CVs will need to be formatted to the specifications of the Terms of Reference. Some TORs allow up to five pages of data for each individual; others allow only two.

- There are various layouts and types of CVs. Refer to specialized books on this topic.

5.6.25   Standards used

Purpose
To describe the standards and codes of practice that will be adhered to.

Guidelines for writing it
List the various standards and codes of practice, together with identification numbers. If you are deviating from the standards requested by the client, you should include a justification of your choices.

5.6.26   Statutory framework

Purpose
To describe the regulations relevant to the project.

Guidelines for writing it
Legal advice should be taken on the wording.

5.6.27   Technical and management skills

Purpose
To describe the technical skills, experience, availability, and commitment of team members.

Guidelines for writing it
Give details of the management skills of team members. Include client liaison, reporting structure, previous performance, qualifications, training and practical experience, and proposed management systems.

5.6.28   Time-based diagrams or charts

Purpose
To describe the interrelationship of tasks required in the work program and to show how this can be achieved in the client's time frame.

Guidelines for writing it
Time-based diagrams or charts are usually presented in Gannt chart format using software such as *Microsoft Project*. They show the start and completion of tasks and use different symbols to identify activities or requirements.

These programs can be updated throughout the course of the project to determine any time-related impacts of the project program. They can then be used to assess how the program can be modified to enable the project to finish on time. It is important to determine the critical path and closely watch it during the course of the project.

Elements that may need to be included are:

- *Staffing schedule* showing when the key personnel will be actively working on the project, linked consistently with the work program.
- *Organization chart* showing coordination and management links (who is in charge on site, who coordinates with the client, and the consultant's home office).
- *Critical path chart* (a standard project management technique): a time-based chart showing which particular tasks are likely to cause delays or disruptions.
- *Man-month estimates table* showing the amount of time each of the key personnel will be working on the project.
- *Data collection.*
- *Design review.*
- *Submission of deliverables* such as the design review report, existing situation report, and the final handover report.

5.6.29   Work program, staffing schedule and organization, *or* Schedule

This can be included instead in Approach and Methodology.

Purpose
To describe the various schedules: organization chart, staffing schedules, critical path chart, man-month estimates, work program, and key personnel tasks. Specifically:

- The tasks, the work areas, and the milestones in the program, and the personnel involved.
- The organization of the project team.
- The project management structure.
- How the liaison and coordination activities will be carried out.
- The resources needed for the project and the base information for the costing of the project.
- How the proposed resources will enable the project to be completed in time.

Guidelines for writing it
▶ Aim for:
  - *Consistency.* Any inconsistencies in your proposal will be noted by the reviewer and cast your company in a bad light.
  - *Plausibility of the program.* You cannot later argue that a better program can be produced. Moreover, some TORs state that the work program as stated in the proposal is legally binding.
  - *Accuracy of resourcing.* It should have the accuracy needed to show that the project team can perform technically, to time and within budget. Inaccuracy will result in either your winning the project or losing it and causing a loss to your company.
  - *Compatibility with the financial proposal.*

▶ Work with project management software for all but the simplest of projects. Changes in the work program arise when personnel continue to revise their ideas. These then cause further changes in other elements of the program. The results are better managed with specialist software.

▶ Present the work program as a series of tables and graphics. The reviewer will more easily understand them in this way. The TOR usually ask for them in this way and often give examples of the layout. Link them together by introductory text, and explain any deviations from the TOR (see Section 5.6.27).

▶ Present key elements of the work program. You will probably need to consider the following elements, even though the Terms of Reference may not ask specifically for all of them. Each element can also be presented as a separate section, as indicated.

• Task lists

• Task sheets. These link the tasks that need to be done to the personnel who will do them.

• Key personnel tasks. These are selected once you know who has to do what. Their CVs can be presented separately in a section *Staff CVs* (see Sections 5.6.15 and 5.6.24).

• Time-based diagrams or charts showing the work program (see Section 5.6.28).

• Staff lists. The personnel who will support the key staff: technical support (local engineers and technicians) and nontechnical support (administrative, guards, etc.).

• Local facilities. Physical assets needed such as local office space, transport, and so on.

Specific Types of Documents

Part 3 gives the requirements for representative types of engineering documents. It should be used in conjunction with Part 2, which gives details of the requirements of the various sections of a document.

It starts by describing the requirements for all types of summaries.

It then gives details of the following types of reports: a major formal proposal, a feasibility study, a due diligence report, an environmental assessment report, a progress report, an incident report, an inspection report, a trip report, a performance review, and a laboratory or research report.

It next describes the requirements for a set of instructions, formal letters, short workplace documents, publicity material, a journal or conference paper, and a conference or display poster.

6

Summarizing

An Executive Summary, a Summary, and a Conference or Journal Paper Abstract

The Executive Summary or Abstract is read by all readers of a document. By giving an initial overview, it plays an important role in explaining the subsequent detail in the document. It is also critical for people who do not need to read the whole document. However, an author's lack of understanding of its significance and requirements can result in inclusion of inadequate material.

This chapter gives information on the different content of the various types of abstracts, together with the requirements for a short summary or abstract of about 200 words, and for the more extensive summary material in an Executive Summary and research paper Abstract.

Structure of the chapter

6.1   Definitions: Executive Summary/Summary/ Abstract

In technical documentation the words "Abstract" and "Summary" are often used interchangeably to mean the same thing. However, the specific differences between the three types of summaries are:

- An *Abstract* presents an overview to an expert audience.
  When required: in specialized documents such as journal papers, conference papers, and posters.

- A *Summary* presents an overview to a less specialized audience. Anyone reading it should be able to gain an understanding of the main features and findings of your document without the detail.
  When required: at the beginning of every document you write if an Abstract or Executive Summary is not specifically requested.

- An *Executive Summary* presents an overview to an executive audience in nonspecialist language. It generally is longer than a Summary, possibly a tenth of the whole document. It should present the work in greater detail than a standard Summary does.
  When required: In a management or consulting document. It may be read by many different types of personnel, some of whom may not have advanced technical expertise. The language therefore needs to be understood by nonexperts.

6.2   The purpose of any type of summary

A Summary should be placed at the beginning of every engineering document—with the possible exception of some types of specifications—for the following reasons:

- To provide the reader with an initial overview. Psychological testing has shown that by first reading an informative overview, a reader can better understand and assess complex information in the rest of the document. Overviews in a document are provided at the two narrow ends of the diamond-shaped structure of a document and by the section summaries (see Chapter 2).
- To give the reader a miniaturized version of the document so that he or she can identify the key information quickly and accurately.
- To help the reader decide whether he or she needs to read the whole document.
- *Journal paper abstract.* To appeal to users of electronic abstracting services. If the Abstract is weak, readers may choose not to access the whole paper.

- *Conference abstract.* To encourage conference organizers to invite you to present a paper.

6.3   Difficulties in writing

It can be difficult to adequately summarize a document in a specified number of words. The difficulties arise from:
- Deciding on the core information.
- Making sure that all aspects are covered.
- Making sure that the abstract does not concentrate on only some aspects at the expense of others.
- Linking the information into a coherent story.
- The final cutting-down process to the stipulated length.
- Not knowing the difference between descriptive and informative abstracts.

6.4   General requirements

▶ **Length of a Summary or Abstract should be proportional to length of the document.**
It should be brief: A lengthy abstract defeats its purpose. As a rough guide to lengths:
- A short document (up to 2,000 words): 200 to 300 words may be enough.
- A relatively long document: 300 words to a page.
- An Executive Summary: 10% or more of the whole document.
- *A conference or journal paper:* The required length will be stated in the Instructions to Authors.

▶ **Aim for an informative, not a descriptive, Summary/Abstract (see below).**

▶ **Do not include any information that does not appear in the main body of the document.**

▶ **Do not use tables, illustrations, or references in a brief Summary/Abstract.**
However, they may be needed in an Executive Summary and in a conference abstract of the longer type (see Section 6.10).

▶ **An Executive Summary can use headings.**
It does not need to be uninterrupted text. Headings are particularly helpful to the reader.

▶ **Section summaries can be useful.**
Larger documents may benefit from having short summaries at the beginning of each section, in addition to the main Summary/Abstract. Section sum-

maries give an overview of the information in that section and are useful navigational tools for the reader. Each one should be headed Section Summary (see Chapter 2).

6.5   Structure

A Summary needs to stand alone. A reader should not need to refer to the rest of the document to understand it. It should not contain intricate details, examples, or explanations. It needs to strictly follow the main line of argument of the document.

The elements of information needed—probably in the order shown—are:

A Summary or Executive Summary in a report
- A statement of the purpose of the report to set the context.
- An outline of the key issues studied.
- An outline of the main conclusion(s).
- An outline of the main recommendation(s), if any.

An Executive Summary in a proposal
All necessary information to enable the tender evaluation team to make a favorable assessment of the proposal. It should as a minimum contain project description, approach, methodology, work program, project team, financial details, deliverables, health and safety issues, quality management systems, and conclusions (critical success factors, the suitability of your organization).

An Abstract in a journal or conference paper
- A statement that places your work in context and describes how your work fits in.
- A brief description of your method of investigation.
- Your main results described briefly and as quantitatively as possible.
- An outline of the main conclusion(s).

6.6   Steps in summarizing

Step 1. Work out the key points.
Ask yourself: "What do the readers most want to know? What questions are they most going to want answered?" (see Section 1.3). It can be useful to list these questions and briefly answer them with key points.

Step 2. Prepare an outline.
Use the answers to these questions.

Step 3. Write the Summary.
Maintain the main line of argument of the report or journal paper.

6.7 The different types of content (descriptive, informative, descriptive/informative)

Based on their content, summaries are generally classified as the following types:

- *Descriptive or indicative.* Avoid writing this type. It gives limited information.
- *Informative.* Aim for this type. It gives real information.
- *Descriptive/Informative.* This type may be necessary in a document such as a literature review or thesis.

Here are examples of descriptive and informative summaries, both constructed around identical material. Note how the descriptive type provides only uninformative material rather than the real information that is needed by a reader, in contrast to the informative summary. Many people write the descriptive type of summary in error. You should actively avoid doing so.

6.7.1 The descriptive or indicative Summary/Abstract

Avoid writing this type.

> #### Example of a descriptive Summary/Abstract
> **Title of document: On-road monitoring of ambient carbon monoxide levels**
> *Abstract*
> This study measures the on-road spatial distribution of levels of carbon monoxide. Methods of measurement are discussed and the difference between on-road and fixed-site data is analyzed. The influence of temperature, wind speed, and humidity is considered. Conclusions as to the effectiveness of this method of carbon monoxide monitoring are given, together with suggested recommendations for future air quality sampling programs.

This describes the structure of the document. It does not give the main findings and conclusions. It is like a Table of Contents.

How to recognize when you are writing a descriptive type of Summary/Abstract

Many people write this type in the mistaken belief that this is what is needed. You can recognize it by the following:

- It describes the structure of the document instead of giving the facts.
- It gives no real information. It does not answer the main questions that a reader would want answered.
- It uses stock phrases that are easily recognized. If you find yourself writing in a Summary any one of the following words or phrases, you can be almost sure that the phrase is describing structure instead of giving real information:

  . . . is analyzed/analyzes

  . . . is considered/considers

> . . . is described/describes
> . . . is discussed/discusses
> . . . is examined/examines
> . . . is presented/presents
> . . . is given/gives

6.7.2 The informative Summary/Abstract

Aim for this type. It gives real information.

Example of an informative Summary

Title of document: On-road monitoring of ambient carbon monoxide levels

A statement to place the investigation in context	This report describes the results of a study using detectors mounted on moving vehicles to measure the on-road distribution of levels of carbon monoxide in Middletown. It compares the results with those obtained from fixed-site monitors. Data from fixed sites have been previously used in air-quality
A statement as to why the study has been done.	monitoring programs; however, there has been doubt about their accuracy in determining levels of carbon monoxide at the adjacent on-road sites.
Method of investigation	Levels of carbon monoxide at 1.5 m above road level were monitored during commuter traffic at peak hours, using a moving vehicle on a selected route where fixed monitors were located. The on-road concentrations were found to be three
The results, quantitatively expressed	times greater than those recorded at the adjacent fixed sites, with mean values of 11.4 ($\pm$ 2.0 SD) ppm and 3.9 ($\pm$ 0.8 SD) ppm, respectively. Levels were also found to increase with decreased temperature and wind speed and increased relative humidity.
The conclusions	It is concluded that fixed-site data are significantly underrepresenting ambient levels and that the methods were effective in measuring the spatial distribution of carbon monoxide, estimating commuter exposure and assessing the effectiveness of fixed-site monitors. An on-road monitoring program is
The recommendation	recommended as a supplement to the present system of monitoring air quality.

Guidelines for writing it

An informative Summary will deliver REAL information to your reader. It describes the purpose of the work, the facts and events, the results, the main conclusion(s), and possibly the main recommendation(s) as briefly and quantitatively as possible.

- ▶ *For an investigative report:* Give specific, quantitative information about methods, results, and conclusions.
- ▶ *For other types of documents:* Give specific information about the topic under investigation, including hard facts and your main conclusions.

▶ *Wording:* Avoid the stock phrases of a descriptive Summary/Abstract (see above). If you find yourself using one of these stock descriptive phrases when you are writing an informative Summary/Abstract, it probably means that a piece of information seems too large to summarize. Reassess it and work out the information that the reader needs.

6.7.3   The Descriptive/Informative Abstract

This is a combination of the two types. It gives specific information about the main results, together with general information about the contents of the rest of the document. A longer document may require this type, but make sure you do not slip into generalizations when your reader needs hard facts.

6.8   An Executive Summary

Please first read Sections 6.4 and 6.7.

Purpose
- To provide a document in miniature that may be read instead of the longer document. It is directed at managerial readers who may not have the appropriate technical knowledge and who may not read the whole report, and at other readers without the necessary expertise.
- *For a proposal:* to give in a succinct overview all the necessary information to enable the tender evaluation team to make a favorable assessment of the proposal.

Length
An Executive Summary may be longer than the conventional Summary/Abstract (apart from some conference abstracts): typically 10% or more of the whole document.

Format
It should be formatted for accessibility of information, and the speed and convenience of the reader. A longer Executive Summary can be organized under descriptive headings and numbered blocks of information, and can be highlighted by boldfacing.

Structure and content
The structure should follow that of the body of the report.

Although the body of the report may contain technical or scientific terminology, the Executive Summary should, as far as possible, be written in nonexpert terms.

For a proposal: It should at a minimum contain project description; approach, methodology, work program, project team, financial details, deliverables, health and safety issues, quality management systems; and conclusions (critical success factors, the suitability of your organization).

6.9   A journal paper Abstract

Please first read Sections 6.4 and Section 6.7.

Many people will read a journal paper Abstract as the original article or in electronic indexing and abstracting services and other secondary publications. Your Abstract needs to be very carefully crafted; from it, people will decide whether it is worth reading the rest of your paper.

Purpose
- To provide a miniaturized version of the whole paper.
- To provide a *brief* summary of each of the main sections of the paper: Introduction; Materials and Methods; Results; and Discussion.
- To enable readers to readily identify the basic content, determine whether it is relevant to their interests, and decide whether they need to read the whole paper.

Guidelines for writing it
▶ Work out the information that a reader doing a database search would like to find in the abstract.

▶ Aim for a fully informative Abstract. Avoid the wording of a descriptive Abstract.

▶ Make sure that the main point of your work is clearly and briefly stated.

▶ Keep the description of the methods in the Abstract brief (unless it is a paper presenting a new method).

▶ Avoid nonstandard abbreviations. Abbreviate a long term in the Abstract only if it is a widely recognized abbreviation. Most journals have a standard list of abbreviations that can be used in abstracts and titles without being spelled out.

▶ Do not cite references in the Abstract.

▶ Keep the maximum length of your Abstract to one paragraph, the requirement for most journals.

▶ Do not exceed the stipulated number of words. Abstracting databases may truncate it, and some of your information will be lost.

Structuring the information
All types of summaries should be clearly structured into three parts: a beginning, a middle, and an end. The beginning—a context-setting statement—is very often omitted in error by engineers; they tend to construct summaries that concentrate on conclusions and recommendations. However, this context statement is important to the readers, particularly those who do not have a detailed knowledge of the topic.

- *The beginning.* One or more statements to put your work in context and to explain the aim of the work. Avoid a general overall statement, for example:

Poor
Carbon monoxide is a major health hazard. This study measures . . .

Better

This report describes the use of detectors mounted on a moving vehicle to measure the on-road distribution of levels of carbon monoxide—a health hazard known to be increasing in Middletown—and compares the levels with those obtained from fixed-site monitors.

- *The middle (Part 1).* A description of the approach you are using. This might be:
  - — An analytical method
  - — A system concept
  - — A scheme of testing
  - — A design technique
  - — A statistical analysis of an environmental problem
  - — Experiments
  - — *[etc.]*

  Make sure that:
  - — You describe the methods in enough detail for the journal or conference reviewers to see that they are valid, but that you leave out all the unimportant detail.
  - — You present the methods in such a way that they can be readily correlated with the results. If you are describing two or more related experiments, make sure that they are clearly demarcated so that the results can be similarly presented.

- *The middle (Part 2).* Your main result. This might be an experimental or theoretical finding or an improved design.

  Make sure that you give as quantitative a description as possible as is consistent with summary length.

- *The end.* The main conclusion(s) or outcome.
  - — Give a brief description of the conclusion(s) you draw from your work or the outcome of it.
  - — *If appropriate,* include your main recommendation(s).

6.10   A conference Abstract

See also Sections 6.4 and 6.7.

Purpose

- *Initial purpose:* To enable the conference organizers to decide whether to invite you to present your work at the conference. There are several aspects to this. They need to decide:
  - — Whether your work fits in with the theme of the conference.
  - — Whether your work is good enough.
  - — If you are invited to the conference, whether you will be asked to present your work orally or as a poster presentation.

- *Purpose during the conference:* To enable each conference participant to decide whether your work is of interest to her or him. They may then:
  — Attend your oral presentation *or* seek out your poster.
  — Try to make personal contact with you during the conference.
  — Contact you after the conference, if they have not met you during it.

Guidelines for writing it
Some conferences ask for a short abstract (100 words to half a page). Others require a two- to three-page self-contained mini-paper. The following points are important:

▶ A particular problem with a conference paper or poster abstract is that it is required months before the conference, when the experimental data may not yet be fully confirmed. One therefore has the problem of devising interesting material before one is sure of the outcome.
▶ Decide on the core material of your work and strip away the detail.
▶ Avoid writing a descriptive Abstract (Section 6.7).
▶ Ensure that your Abstract, like a short story, has a beginning, a middle, and an end. These parts need to be clearly defined and in the correct order. Use the same scheme for the beginning, middle, and end as for a journal paper Abstract (see Section 6.9).
▶ *Short Abstract:* Do not include illustrations or references.
▶ *Long Abstract:* Do not use more than one or two small figures or tables. You will need to restrict the number of references so that the References section does not take up too much of your limited space.

Possible headings
- *Short Abstract:* No headings.
- *Long Abstract:* A basic skeleton of headings required is often stipulated. It may resemble the standard structure of a journal paper (Abstract, Introduction, Methods, Results, and Discussion. See Chapter 12.)

Formatting and appearance
Most conferences require Abstracts to be submitted as "camera-ready" copy; that is, when they are reproduced, they will look exactly as they do when you submit them.

Conference organizers strive for a uniform appearance; the *Instructions to Authors* will give precise requirements for font and type size, margin size, and so on. Be sure that the *Instructions to Authors* are followed in every detail, even if you think you can improve the appearance by using a different font or size from those that are stipulated. If instructions are not followed, the Abstract probably will be returned to you for alteration.

6.11   Common mistakes in Abstracts or Summaries

- No clear statement of the main point of the document.

- Too descriptive: describes only the structure of the document, gives no real information.
- Too long and too detailed.
- Conversely—in the attempt to cut it down to the required number of words—large editorial inconsistencies (usually gaps in the logical flow of the information).
- Vague, imprecise information.
- An illogically presented story.
- Important information missing, e.g., a clear indication of the methods used, a quantitative description of the results.
- No clear conclusions.
- Unfamiliar abbreviations.

Checklist for Executive Summary/Summary/Abstract

Does it include the following (if appropriate to the subject matter):
  ☐ A statement that places your work in context?
  ☐ Your method of investigation?
  ☐ Your main results or observations?
  ☐ Your main conclusion(s)?
  ☐ Your main recommendations(s)?
☐ If these are inappropriate to your subject matter: does it give *real* information?
☐ Have you avoided writing a descriptive type of Summary/Abstract?
☐ Is there any information that does not appear elsewhere in the document? If so, incorporate it somewhere.
☐ Did you, in keeping with the general format for a Summary/Abstract, omit tables, figures, literature references? (except for a conference abstract, see below).
☐ Was the Summary/Abstract written at the final stage of the document?

An Executive Summary
☐ Does it give the information appropriate to a managerial readership?
☐ Is it written so that a nontechnical person can understand it?
☐ Does it follow the structure of the main document?
☐ Does it have appropriate descriptive headings and numbered blocks of information, and is it highlighted by boldfacing?
☐ Is it written for accessibility of information and the speed and convenience of the reader?
☐ *For a proposal:* Does it contain project description, approach, methodology, work program, project team, financial details, deliverables, health and safety issues, quality management systems, and conclusions (critical success factors, the suitability of your organization)?

A journal paper Abstract

☐ Is your conclusion clearly stated?

☐ Is the story presented logically?

☐ Does it have a beginning (the context), a middle (methods and results), and an end (the conclusions or outcome)?

☐ Does it contain the sort of information that a reader doing a database search would like to find?

☐ Is the description of the methods brief (unless the paper is presenting a new method)?

☐ Do the results make up most of it?

☐ Have nonstandard abbreviations been avoided?

☐ Have citations been avoided?

☐ Have you kept to the word limit? If not, databases may truncate it.

A conference Abstract

☐ Does it conform to the conference guidelines (page number, etc.)?

☐ Have you included only one or two small illustrations?

☐ Does the story have a beginning (the context), a middle (methods and results), and an end (the conclusions or outcome)?

☐ Have you included an appropriate number of citations and a short List of References?

7

Reports

This chapter gives information on the major types of reports written by civil engineers. It describes the requirements for each of the specific types of reports and presents a suggested structure of section headings for each one. Each structure is cross-referenced to the requirements of the various sections as described in Chapter 5. The proposed structures are suggestions only; they should be modified as necessary.

Examples of complete documents are not given. Most reports written by civil engineers are large documents; it would trivialize them to try to present short representative versions. For a description of the basic structure of a generalized engineering report, the reader is referred to Chapter 2.

The chapter does not cover the writing of specifications. They can be complex, the language needs to be unambiguous, and errors can be critical; readers are therefore referred to specialized sources on this subject (see Part 7).

Structure of the chapter

7.1 Major formal proposal
7.2 Feasibility study
7.3 Due diligence report
7.4 Environmental assessment report
7.5 Progress report
7.6 Incident report
7.7 Inspection report
7.8 Trip report
7.9 Performance review
7.10 Laboratory or research report

7.1   Major formal proposal

Note: This section's material can also be applied to a Capability Statement.

Most formal proposals are usually written in response to a Request for Proposal (RFP). There usually is little time between the issue of the RFP and the submission date. Since large teams usually work on putting together a proposal, it is critical to coordinate everyone's efforts to the greatest effect.

An RFP usually calls for both a technical proposal and a financial proposal. This chapter deals only with putting together a technical proposal; for preparing financial proposals, please refer to specialized texts.

7.1.1   Guidelines for writing it

Using stock material
Most companies have had experience putting together large proposals and will have a collection of old documents that will have been both successful and unsuccessful in winning contracts. It is tempting to cut and paste parts of this material in part-production of the new proposal. The dangers associated with doing this are:
- New recruits to the writing process often do not know if the material they are copying is a good example to use.
- The cutting-and-pasting process can introduce errors of copying and insertion.
- The client may already be familiar with the material from old proposals and may see your reuse of it as sloppiness.

To avoid these dangers, an organization may find it useful to keep a library set of proposals that are regarded as good examples of their type in their content, formatting, and style of writing.

Presentation style
Official evaluation systems of proposals tend to award only about 2% of the total points to the quality of presentation. However, readers are only human; if any document makes a good first impression and if the information can be easily found, then it tends to be more highly rated overall. Where the technical content of two competing proposals is comparable, the better-presented proposal will win.

For page layout and orientation, font and type size, and binding, see Chapter 4. For page and section numbering and other conventions, see Chapter 15.

A suggested structure for a proposal
Although there is some similarity between the structures set forth for RFPs by the various organizations and government agencies, no standard format can be recommended here. *However, it is vitally important to provide all the material*

that the RFP asks for and in the same order as requested. Do not be tempted to alter the order because you believe you can improve on it.

Here is a basic plan commonly used in a generalized form for a large technical proposal. It has a basic structure that can be divided in simple terms into the first, core, and final parts. The section headings that are here shown associated with these three parts are those that are commonly found in large proposals.

For full details about the sections, see Chapter 5.

The first part (the Executive Summary and the sections that explain the purpose and contents of the document):
- ▶ **Title page** (edge page, if using a lever-arch binder).
- ▶ **Letter of transmittal.**
- ▶ **Executive Summary.** Succinct overview of whole document: all necessary information to enable the tender evaluation team to make a favorable assessment of the proposal. Should as a minimum contain project description, approach, methodology, work program, project team, financial details, deliverables, OSH and quality management systems, conclusions (critical success factors, suitability of your organization).
- ▶ **Table of Contents.**
- ▶ **List of Illustrations.**
- ▶ **Glossary of Terms.**
- ▶ **Introduction** *or* **Background** *or* **both.**

The core part (the sections that are the most important in a proposal's evaluation or formal marking):
- ▶ **Company Description and Experience.** Description of the organization's/consortium's general capability and experience. If appropriate, can be split into two separate sections: **Company Description** and **Company Experience.**
- ▶ **Site Appreciation.** Description of site, technical notes, photographs.
- ▶ **Project Understanding and Objectives.** Outlines how the project fits into the client's overall program and provides more background information to the proposed approach and methodology.
- ▶ **Master schedule/master plan.**
- ▶ **Approach and Methodology.** How you intend to execute the project. This section could have three elements: List of Tasks, Technical Methods, Key Issues.
- ▶ **Work Program, Staffing Schedule and Organization.** Organization chart, staffing schedules, critical path chart, man-month estimates, work program, key personnel tasks.
- ▶ **Time-Based Diagrams and Charts.**
- ▶ **Comments on the TOR.** Summary of your proposal with respect to the Terms of Reference, additional tasks and optional tasks, ambiguities or inconsistencies in the Terms of Reference, potential problems that may not have been noted in the Terms of Reference.

- ▶ **Staff CVs.** Current CV (biodata) for each staff member involved in the project.
- ▶ **Other sections** of your choice, e.g., Health and safety issues, Intellectual property/Client ownership, Quality assurance, and so on.

The final part (the additional, necessary sections that do not fit into the beginning or the core):

- ▶ **Association Arrangements.** Arrangements for association with other companies, initialed copy of TOR, signed copy of other documents.
- ▶ **Local Facilities and Counterpart Staff.** Describes the counterpart office and support requirements, e.g., local staff, accommodation, equipment, and information, and identifies who will provide them.
- ▶ **Appendixes** *or* **Attachments.**

7.1.2   Monitoring progress of the report

- ▶ *Keep a master copy of the proposal.* This will mean that everyone will have a central place in which to follow its progress and get ideas for its improvement, its overall form can be readily monitored, and actual and potential omissions and errors can be noticed.
- ▶ *Have a central system for storing computer files,* if not a huge centralized system covering all the work of the organization, at least for each separate proposal. This will mean that there is ready access to such things as general correspondence files, text files, graphics, staff CVs, e-mail files, and contact addresses.
- ▶ *Encourage team leaders to keep a project diary.* This will help the recall of decisions, promises, and allocation of duties, all of which may be difficult to remember later.

7.1.3   Final quality assurance of the completed proposal

A proposal needs to have two types of quality assurance (QA) checks: the technical QA check and the style QA check.

- ▶ The *technical QA* checks the content. The document must be checked by one or more experienced senior engineers to ensure that the technical argument is convincing, error-free, and consistent and that it complies with all technical requirements of the RFP.

- ▶ The *style QA* checks the document integrity and appearance for the following:
  - Consistency of font and formatting
  - Numbering of sections, pages, illustrations
  - Correspondence between numbering in the text and in the Table of Contents
  - Headers and footers
  - Page breaks
  - Style of writing, particularly when many authors have contributed

7.1.4   How proposals are marked or rated

According to Bartlett,* the sections of a technical proposal that count most toward its evaluation are:

- The site and project appreciations
- The approach and methodology
- Task lists
- The work program and staffing schedule
- The qualities of the staff and their CVs.

An organization's proposal effort therefore needs to concentrate on these sections.

7.2   Feasibility study

A feasibility study considers a range of factors (plant size, siting, etc.) or designs to establish whether something can be feasibly constructed. It is essential that outcomes meet clients' expectations.

A suggested structure for a feasibility study

- ▶ **Title page.**
- ▶ **Letter of transmittal** (Cover letter), if needed.
- ▶ **Executive Summary.** Brief description of the purpose of the study, the key conclusions and recommendations, a statement of the total estimated cost and project activity time.
- ▶ **Table of Contents.**
- ▶ **List of Illustrations.**
- ▶ **Acknowledgments.**
- ▶ **Glossary of Terms** and **Abbreviations.**
- ▶ **Introduction** *or* **Background** *or* **both.**
- ▶ **Scope.** A statement of the purpose of the report and the scope of work for the study, together with all the user requirements that must be satisfied.
- ▶ **References.** Identify all documents that relate to the study.
- ▶ **Objectives.** A list of objectives for the study. These could include such things as minimize demands on existing services, minimize installation and set-to-work costs, maximize the use of commercially available components, fully meet the requirements of specific standards, maximize ease of use for system operators, and so on.
- ▶ **Options.** This forms the body of the report and contains a technical evaluation of the options being considered. Subheadings and functional descriptions, as well as system and subsystem block diagrams, should be used where appropriate. Justification should be provided for

---

*Bartlett, R.E. (1997) *Preparing International Proposals*. Thomas Telford, London.

options that are recommended for implementation. A list of any equipment added, modified, or removed should also be included.

▶ Other sections relevant to the project could include:
  • Present situation
  • Environmental issues
  • Cost estimates
  • Project management, including scope of work, cost, schedule, risk management, and assumptions

▶ **Impact Analysis.** Statements of the impact that the project or design may have on specified areas.

▶ **Conclusions.**

▶ **Recommendations** or one combined section: Conclusions and Recommendations.

▶ **References** (if needed).

▶ **Appendixes** *or* **Attachments.**

7.3   Due diligence report

A due diligence report must show that the diligence due to the process of the investigation is being observed. There is no checklist that covers all possible situations. Rather, what and how much diligence is due is measured by a standard of reasonableness and prudence under the particular circumstances. On the other hand, if the cost of conducting a particular test or taking a particular precaution significantly outweighs the likely resulting benefits, due diligence does not require that it be conducted or undertaken.

A due diligence report may involve examining issues such as:
  • The condition and value of assets
  • Environmental issues
  • Geotechnical and geological issues
  • Safety issues

A suggested structure for a due diligence report

▶ **Title page.**

▶ **Letter of transmittal** (Cover letter), if needed.

▶ **Executive Summary.** Brief description of the purpose of the study, the key conclusions and recommendations, a statement of the total estimated cost and project activity time.

▶ **Table of Contents.**

▶ **List of Illustrations.**

▶ **Acknowledgments.**

▶ **Glossary of Terms** and **Abbreviations.**

▶ **Objectives.**

▶ **Scope.**

▶ **Quality Assurance.**

- ▶ **Introduction.** Including, e.g.
  - Initiation of project
  - Extent of assets studied
  - Other parties involved
  - Site inspections
  - Drawings/photographs/life of assets
  - Description of any databases (e.g., on CD) included as part of the report
  - What the report is *not* covering
- ▶ **Company Description and Experience.** Material directly relevant to due diligence procedures.
- ▶ **Assessment of various aspects.** Including, e.g.
  - Environmental aspects
  - Assets and asset management
  - Structural engineering assessment
  - Geotechnical/geological aspects
  - Safety audit
  - Valuation
  - Cost/capex issues (financial implications)
  - Other issues (followed by a series of sections appropriate to the subject matter)
- ▶ **Standards.**
- ▶ **Conclusions.**
- ▶ **Recommendations** (or one combined section: Conclusions and Recommendations).
- ▶ **References** (if needed). *Note:* The auditor or reviewer will look for rigorous management of information. Statements need to be robust and able to be sourced, and therefore well referenced.
- ▶ **Appendixes** *or* **Attachments.**

7.4   Environmental assessment report

An environmental assessment report investigates a situation from the environmental standpoint. It can also possibly describe and assess the risks to human and environmental health and propose remediation measures.

A suggested structure for an environmental assessment report
- ▶ **Title page.**
- ▶ **Letter of transmittal** (Cover letter), if needed.
- ▶ **Executive Summary.**
- ▶ **Table of Contents.**
- ▶ **List of Illustrations.**
- ▶ **Acknowledgments.**
- ▶ **Glossary of Terms** and **Abbreviations.**
- ▶ **Objectives.**

- ▶ **Introduction** *or* **Background.**
- ▶ Possible other sections required:
  - Site Appreciation.
  - Statutory Framework.
  - Records Review.
  - Environmental investigation/methodology used, including any software used for environmental modeling.
  - Risk Assessment.
  - Other sections relevant to the project.
- ▶ **Conclusions.**
- ▶ **Recommendations** (or one combined section: Conclusions and Recommendations).
- ▶ **References** (if needed).
- ▶ **Appendixes.**

7.5   Progress report

A progress report keeps others informed about your or your team's progress, with the objective of project monitoring and accountability. The projects can vary in length from a few weeks to several years; project reports that describe them are submitted at intervals, often as specified in the original project proposal.

The general format can vary from an informal memo or letter giving an update on work to a very detailed formal structure required by the client or funding body.

It is essential that each progress report should, in the main, be self-standing.

A suggested structure for a progress report
- ▶ **Title page.**
- ▶ **Letter of transmittal** (Cover letter), if needed.
- ▶ **Executive Summary.**
- ▶ **Table of Contents.**
- ▶ **List of Illustrations.**
- ▶ **Acknowledgments.**
- ▶ **Glossary of Terms** and **Abbreviations.**
- ▶ **Objectives.**
- ▶ **Introduction** *or* **Background** *or* **Project Summary.**
  - History: Description of the events leading up to the present situation; what had been planned.
  - *If it is to be circulated to people who are unfamiliar with the project:* Description of the people involved, the location of the project, and the dates.
- ▶ **Planned Work.** A description of what should have been completed by the reporting date (with details in an attachment).

- ▶ **Progress.** A description of how much work has been completed.
  - • *If work has been completed to schedule:* Only brief details.
  - • *If there have been deviations from the plan:* Explain the problems that led to these variations being necessary, and refer the reader to the following section, Problems Encountered.
- ▶ **Problems Encountered.** Describe in detail any problems (cost or schedule issues); actions taken to deal with them; the success of those actions; the effect on the costs or schedule.
- ▶ **Future Work.** Work plans; cost or schedule adjustment, if any, with reasons (refer the reader back to the previous section).
- ▶ **Adherence to Schedule.** Description of the original schedule. This is necessary even if it is unaltered from the previous progress report. State whether the project is on schedule, ahead of schedule, or behind schedule.
- ▶ **Predicted Schedule.** If appropriate, present an updated schedule to account for alterations in planning.
- ▶ **Conclusions.** Brief overview of whether the project is running to costs and to schedule and any predicted alterations.
- ▶ **References.** (if needed)
- ▶ **Appendixes** *or* **Attachments.** Detailed schedules, costings, drawings, etc. If there have been deviations from the plan, explain in detail the problems that led to these variations being necessary, how they have been dealt with, and effects on schedule and costs.

7.6 Incident report

An incident report describes an event that has happened, how and why it occurred, the effect of the event, and actions that have been taken (if appropriate) and further actions that may need to be taken.

A suggested structure for an incident report

- ▶ **Title.**
- ▶ **Summary.** A brief overview of the event, its outcome, and actions taken.
- ▶ **Introduction** *or* **Background.** The circumstances leading to the event.
- ▶ **The event.** A description of what happened.
- ▶ **Outcome.** The effect of the event; action that has been taken.
- ▶ **Conclusions.** Brief description of the end result; the effectiveness of the actions.
- ▶ **Recommendations** (or one combined section: Conclusions and Recommendations). List of the actions that may still need to be taken.
- ▶ **Appendixes** *or* **Attachments.**

7.7   Inspection report

An inspection report describes an inspection that has been made, for example, of construction work.

A suggested structure for an inspection report
- ▸ **Title.** Informative title.
- ▸ **Summary.** Brief overview of what you found out, the implications, actions to be taken, and by whom.
- ▸ **Introduction** *or* **Background.** Circumstances behind the inspection: why it was necessary, and its implications.
- ▸ **Present Condition.** Description of the condition of the items of the facility that you inspected.
- ▸ **Deficiencies.** Conditions that need to be corrected.
- ▸ **Work Required.** Actions to be taken and by whom.
- ▸ **Conclusions.**
- ▸ **Recommendations** (or one combined section: Conclusions and Recommendations).
- ▸ **Appendixes** *or* **Attachments.**

7.8   Trip report

A trip report describes a job that you have done away from your usual place of work (e.g., installation, modification, or repair of equipment; field project; conference or workshop; evaluations). It should include the reason for the trip, what was found, and your conclusions.

A suggested structure for a trip report
- ▸ **Title.**
- ▸ **Summary.** Brief overview of purpose of trip, what was done, or what was found out.
- ▸ **Background** *or* **Purpose of the trip.** The purpose of the trip, who went, why, and when.
- ▸ **The job or event.** Description of what was done, how it was done, and what was found out.
- ▸ **The result.** The result of what you did, its effects, and what else needs to be done.
- ▸ **Conclusions.**
- ▸ **Recommendations** (or one combined section: Conclusions and Recommendations).
- ▸ **Appendixes** *or* **Attachments.**

7.9 Performance review

A performance review reports on the performance of a system and assesses its efficiency.

A suggested structure for a performance review
- ► Title page.
- ► **Letter of transmittal** (Cover letter), if needed.
- ► **Executive Summary.**
- ► **Table of Contents.**
- ► **List of Illustrations.**
- ► **Glossary of Terms** and **Abbreviations.**
- ► **Objectives.**
- ► **Introduction** *or* **Background.**
- ► **Methodology** *or* **Testing procedures.**
- ► **Results** *or* **Outcome.**
- ► **Conclusions.**
- ► **Recommendations** (or one combined section: Conclusions and Recommendations).
- ► **Appendixes** *or* **Attachments.**

7.10 Laboratory or research report

A laboratory or research report describes investigations into properties and principles of objects, systems, or concepts. It has the main aims of giving enough detail of the procedure to enable it to be repeated and verified by another competent person and to present the results and draw conclusions from the work.

A suggested structure for a laboratory or research report, based on the AIMRAD format
The structure of these reports is based on the simple Abstract, Introduction, Methods, Results, and Discussion (AIMRAD) concept. These headings in themselves may be appropriate for the report. However, it may be necessary to devise different section headings while remaining within this broad structure of giving the background to the work, describing the methods, and presenting the results and the conclusions you draw from them.
- ► Title page.
- ► **Executive Summary** *or* **Abstract.**
- ► **Table of Contents.**
- ► **List of Illustrations.**
- ► **Glossary of Terms** and **Abbreviations.**
- ► **Objectives** *or* **Aim.**
- ► **Introduction** *or* **Background.**

- ▶ **Materials and Methods** *or* **Procedure.**
- ▶ **Results.**
- ▶ **Discussion** (or a combination of the two preceding sections: Results and Discussion).
- ▶ **Conclusions.**
- ▶ **Recommendations** (or one combined section: Conclusions and Recommendations).
- ▶ **References.**
- ▶ **Appendixes.**

8

A Set of Instructions
Handbook, Procedure, Operating Manual

A user manual, hard copy or electronic, enables someone to interact with an item of technology safely and effectively. It gives instructions on how to assemble, operate, maintain, adjust, repair, modify, or troubleshoot something. It is therefore an item for training inexperienced users and instructing repair and installation personnel.

When writing a user manual, the aim should be to write precise and accurate instructions so that others can understand them. It is vital to be able to place yourself in the users' position and judge what you have written from their standpoint, not from your own set of expert skills. The language used should be concise and capable of only one interpretation.

This chapter suggests a possible structure for a procedure and gives guidelines for the unambiguous wording of the instructions. It covers the requirements for a relatively simple set of instructions. For procedures that carry critical levels of risk, please refer to specialized books on procedure writing.

Structure of the chapter

8.1   Aim

To write instructions that will enable people to clearly interpret your meaning and to act safely and effectively when putting them in place.

8.2   Difficulties

Knowing your readership and under what circumstances people will be using the product or system you are describing. This type of "how-to" writing is often confusing and poorly written. It comes from not realizing that most other people do not have the in-depth knowledge of the system that you have. To avoid this, place yourself in the reader's mind and work out what the reader needs to hear from you. Distance yourself from your own knowledge.

8.3   Possible structure for a procedure

- **A short purpose statement.**
Many writers assume that since they are telling readers what to do, there is no point in telling them why to do it. This is dangerously wrong; most people will not automatically see the wisdom of doing it your way or the possible dangers involved. Include (a) an introduction to the material, explaining the purpose for and the importance of the instructions; (b) a brief overview of how the product or system works; (c) why the instructions must be followed; (d) what will be achieved.

- **Tools and materials required**
List the tools and the materials that the reader will need to assemble.

- **Glossary of Terms and Abbreviations**
Give clear, precise definitions. Place this section at the front of the document so that the reader can easily find it (see Section 5.3.7).

- **Special instructions, e.g., safety warnings**
Prominently display any special items such as safety warnings. But make sure that an important warning is also repeated in the instruction to which it relates.

- **Series of instructions**
Give a numbered series of instructions according to the guidelines in the next section.

8.4   Guidelines for wording of the instructions

- ▶ **Use headings that have widely accepted meanings:**
    - *Danger:* Reserved for steps in a procedure that could lead to injury or loss of life.
    - *Warning:* Used for steps that could result in damage to the product.
    - *Caution:* Used where faulty results could occur.

- *Comment:* Used to alert the reader to a potential problem and to make suggestions that would make the reader's task easier.

▶ Remember that most readers carry out each instruction after reading it, without knowing what comes next

This can have critical implications when writing whole sets of instructions and individual instructions. Some of the following guidelines illustrate this.

▶ Use simple words

Use the simplest words possible. Do not use jargon—for example, do not write "deactivate" when you mean "turn off."

▶ Restrict the use of abbreviations and acronyms

It is possible for abbreviations to be less widely understood than a writer of instructions may think. It is important to consider whether you really need to use an abbreviated form. For any that you do need to use, explain them in a Glossary of Terms and Abbreviations placed at the beginning of the set of instructions (see Section 5.3.7).

▶ Use sentences that are as short as possible

Long and complex sentences can be misinterpreted.

▶ Be clear and unambiguous

Terms such as "front" and "back" or "left" and "right" can be confusing. If you were placed in front of the device when writing the instructions, left and right are reversed for the repair technician standing at the back. And it might be the right way up when installed, but upside down when being repaired. Therefore avoid these terms whenever possible or carefully explain the viewing direction.

Ineffective

> Make sure the switch is in the upward position, then close the drain valve.

Effective

> Make sure the switch is in the OFF position, then close the drain valve.

▶ Write in numbered steps

Steps are numbered to indicate how sections, steps, and sub-steps are related. The step-numbering scheme should not be too complex for the procedure and should not be too complexly subdivided.

▶ Use the imperative form of the verb—one that gives an instruction

Give orders clearly so that there is no mistaking what you mean. Avoid the word *should*.

Ineffective

> The power switch should be turned off.
> You should turn the power switch off.

Effective

> Turn the power switch to OFF.

Negative instructions are also effective
> Do NOT turn the activator dial.

▶ Let each instruction require only one action

The possibility for confusion is reduced if you make sure that only one action is contained in each numbered instruction.

Ineffective
> Ensure that both the water supply valve and the feed valve are open, then start the transfer pump by pressing the START button.

Effective
> 1. Ensure that both the water supply valve and the feed valve are open.
> 2. Start the transfer pump by pressing the START button.

▶ Avoid negative statements

"Ensure that the valve is not open" is a negative statement. "Ensure that the valve is closed" is a positive statement and is preferable. If a negative statement is followed by a "then" clause, confusion can result. Avoid statements like "IF the valve is not open, THEN do not adjust flow."

▶ Write one-way directions

The following instructions are in the reverse order. Obeying them in the sequence given could ruin the pump and probably damage the preset control valve.

Poor example
> 1. Start the pump.
> 2. Before starting the pump, check to see that the cooling-water valves are open and that the control valve is open.
> 3. The control valve is preset and should not be adjusted.

Rewritten as one-way directions
> 1. Make sure that the control valve is open. Do not adjust it; it is preset.
> 2. Make sure that the cooling-water valves are open.
> 3. Start the pump.

▶ Use conditional procedures effectively: "If … then" and "If … when"

If the procedures are conditional on something (if or when something happens), then say so at the beginning. Do not give the instruction first.

- *Using "if … then."* The pattern "if … then" asks the reader to consider whether the condition applies before carrying out the action. "If" and "then" should be emphasized by capital letters.

Ineffective
> Switch off the pump IF the valve does not open.
> DO NOT carry out the following procedures IF the temperature is above 75°F.

Effective
> IF the valve does not open, THEN switch off the pump.
> IF the temperature is above 75°F, THEN DO NOT carry out the following procedures.

- *Using "if ... when."* Using *when* indicates something that is expected to occur while a procedure is being carried out. The format for "when ... then" is the same as for "if ... then."

  Ineffective
  > Open the valve WHEN the temperature is below 75°F.

  Effective
  > WHEN the temperature is below 75°F, open the valve.

▶ **Do not leave important actions to the discretion of the reader**

Avoid words such as "should" or "may."

  Poor example
  > The line may need to be drained.

  Rewritten so that reader does not have to use discretion
  > 1. Read and record the level on the sight glass.
  > 2. If the level is greater than 10 cm, open Valve D.
  > 3. Drain the line.
  > 4. Close the valve when steam begins to come out of the valve.

▶ **Do not leave out vital information**

Remember that you are familiar with the procedure; your reader is not. Do not assume that the reader will understand what you meant to say. State it explicitly so that the reader does not have to think, only to act.

▶ **Differentiate between the use of figures (Arabic) and spelled-out numbers in a consistent way**

See Section 15.1.

▶ **Specify ranges rather than error margins**

A range of values shows the extent of the acceptable values (e.g., 18–22). Error margins or bands show the acceptable deviation from a number (e.g., 20 ± 2).

Error bands require a user to make a mental calculation; this may allow error to creep in.

  Ineffective
  > IF the temperature is 75°F (±5°F), THEN open the valve.

  Effective
  > IF the temperature is 70–80°F, THEN open the valve.

▶ **Use effective formats: ragged right margins, white space, place-keeping aids**

Use ragged right margins; justified text is more difficult to read than ragged right because of the variable spacing between the words.

To reflect the step structure of the instructions, create white space by indenting, using bullet (dot) points, using space above and below steps and headings, and so on.

Use place-keeping aids: These are boxes or lines that users check as they perform a step.

▶ Use emphasis techniques in a consistent way to highlight information

Emphasis techniques can include using **boldface**, *italics*, <u>underlining</u>, or ALL CAPITALS.

However, avoid using them for large blocks of text; they are difficult to read. Also, use them consistently and with restraint; too much emphasis will kill the original intention.

▶ Provide warnings and cautions

- *Safety instructions and warnings must be prominently placed at the beginning of the instructions.* Do not place cautions or warnings at the end of an instruction.

Poor examples

The main point of this message—that there is a danger of flashback—is at the end.

Valve X is not to be opened before cooling to 75°F because of the possibility of flashback.

Do not open Valve X before it has cooled to 75°F; there is a danger of flashback.

On no account should you open Valve X before it cools to 75°F or you may cause flashback.

Rewritten

The main message is now at the beginning, highlighted by the word "Danger." This is followed by a strong negative instruction.

Danger of FLASHBACK: do not open Valve X before it cools to 75°F.

- *If there is a safety aspect, give the warning first.* Do not give the instruction first.

Ineffective

Light the match and slowly bring it toward the nozzle. Do not light the match directly over the nozzle: It may cause an explosion.

Effective

WARNING: Do not light the match directly over the nozzle: It may cause an explosion. Light the match and slowly bring it toward the nozzle.

Checklist for a procedure or set of instructions

☐ Did you use the words "danger," "warning," "caution," and "comment" in accordance with their widely accepted meanings?

If there is a safety aspect:

☐ Are safety instructions and warnings prominent at the beginning of the instruction?

☐ Is the warning given first, before the instruction?

☐ Did you list the tools and materials required?

☐ Is there a Glossary of Terms and Abbreviations that gives clear, precise definitions?

☐ Are special instructions such as safety warnings prominently displayed?

☐ Have you considered that many people carry out each of the instructions while reading it, without knowing what comes next?

☐ Is the imperative form of the verb used for each instruction?

☐ Did you use simple words?

☐ Did you use short sentences?

☐ Did you avoid jargon?

☐ Does each instruction require only one action?

☐ Did you restrict the use of abbreviations and acronyms?

☐ Does each instruction require only one action?

☐ Did you avoid negative statements?

☐ Is each direction one way?

☐ If the instruction is conditional on something, is the pattern "if ... then" used?

☐ Have all important actions been clearly explained?

☐ Did you include all vital information?

☐ Is each instruction clear and unambiguous?

☐ AGAIN: Is the wording simple? Is the whole procedure safety-conscious?

9

Formal Letters

Letters used in formal or semiformal professional contexts provide a hard copy record, which can be important in a legal context. They are more formal and permanent than e-mails and faxes.

This chapter covers the elements relevant to the formal types of letters written by engineering organizations and gives guidelines for letters that accompany a document.

Structure of the chapter

9.1   The conventions: the elements of a formal letter

The components of a formal letter need to be arranged in a particular sequence, as dictated by the conventions of formal letter writing.

All elements are left justified, except for the subject heading.

9.1.1 Summary of the individual elements

- Your address or institution's address (or letterhead)
  2-line space
- The date
  2-line space
- Name and address of the person to whom you are writing
  3-line space
- The greeting (salutation)
  2-line space
- The subject heading
  2-line space
- The body of the letter
  2-line space
- The closing
  Leave a 6–8-line space for your signature
- Your written signature with your typed name and position below it
  2-line space
- The letters "Encl." if you are enclosing additional documentation with the letter

9.1.2 Details of the individual elements

▶ **Your address or institution's address (or letterhead)**
- The address should be left justified. (Note: The conventions of a few years ago dictated that it should be right justified. This is now regarded as old-fashioned.)
- If you are using letterhead paper, an address is not needed.
- It is now no longer the convention to put a comma at the end of each line of the address.

▶ **The date**
Use the format: *Month* (written out) *Day* (in figures) *Year in figures* (with a comma between the day and the year).

Correct
July 8, 2006

Incorrect
8/7/06 *(different countries use different formats when using only figures; it can cause confusion)*
8 July 2006 *(regarded as old-fashioned)*
July 8th, 2006

▶ **Name and mailing address of the person to whom you are writing**
- The name and address should be left justified.
- Commas are not needed at the end of each line.

▶ **The greeting (salutation)**

According to the tone of the letter, choose from:

- *Dear Sir* or *Dear Madam:* Use in formal situations, either when you do not know the family name of the person or when it would be inappropriate to use it.
- *Dear Sir/Madam:* Use in formal situations when you do not know the family name or the gender of the person to whom you are writing.
- *Dear Mr. [surname].*
- *Dear Mrs. [surname].*
- *Dear Ms. [surname]:* Use when the marital status of the woman is not known or has not been specified.
- *Dear Dr. [surname].*
- *Dear Prof. [surname].*
- *Dear [first name]:* Use when you are on familiar terms with the person to whom you are writing, but still need to use a formal letter format.

Include abbreviations of degrees, certifications, and affiliations, if known, following the addressee's name:

- John M. Smith, Ph.D.; Dear Dr. Smith
- Mr. John M. Smith, P.E.; Dear Mr. Smith
- John M. Smith, Ph.D., P.E., F.ASCE; Dear Dr. Smith

▶ **The subject heading (title)**

- A concise title, two lines below the greeting, centered, boldfaced for emphasis. It should give the reader instant access to the main point of the letter.
- Do not use "Re:" before the title. It is meaningless and old-fashioned.
- Do not underline—this is old-fashioned. Use boldfacing.

▶ **The body of the letter**

See Section 9.3.

▶ **The closing**

Classic letter writing conventions dictate the following:

- If you have used "Dear Sir," "Dear Madam," or "Dear Sir/Madam," then you should use "Yours faithfully," as the closing.
- If you have used the surname in the salutation, then you should use "Yours sincerely," as the closing.

This rigid convention has now been considerably relaxed. Many companies now favor "Yours sincerely," whatever the initial salutation.

- If the letter is not strictly formal, the tone of the letter can be softened by using "Regards" or "Kind regards" or "Best wishes," either before the closing or alone.

▶ **Your written signature and your typed name and position below it**

- After the closing, leave about eight blank lines for your written signature.
- Then provide (left justified) your name. It is becoming good practice to use your full first name and surname (e.g., Joe Bloggs), not initials and

surname (not J. F. Bloggs). This can convey to the reader your gender—making you easier to contact—and gives the sense of a real person behind the letter.

- On a new line, provide your position in the organization.

▶ **The letters "Encl." if you are enclosing additional documentation with the letter**

9.2 Font, spacing, arrangement on the page

▶ Choose a simple serif (e.g., Times Roman) or sans serif (e.g., Arial) font. Elaborate fonts are more difficult to read and give the wrong impression.

▶ Use 10- or 12-point type size.

▶ In the text of the letter, use single line spacing and leave one blank line between paragraphs.

▶ Use ample margins.

▶ If the letter is short, adjust the various spacings so that it is not squashed into the top part of the page.

▶ Second and subsequent pages should include a header that contains the addressee's name, the date, and the page number, on separate lines.

▶ The last page should not contain only the signature. If necessary, reduce the type size.

▶ All the elements should make a pleasing arrangement on the page.

9.3 Structure of the information

▶ **Main point at the beginning: supporting information after the main point**

- *First paragraph:* (If appropriate) a courteous acknowledgment of letter, phone call, and so on.
- *Second paragraph:* The main point. *This is important:* Do not lead up to the disclosure at the end; start with it. It applies to all letters, including those that convey bad news.
- *The following paragraphs:* The information supporting the main point.
- *Last paragraph:* Be careful not to finish abruptly. A courteous final paragraph is needed (e.g., "If you have any further queries, please do not hesitate to contact me.")

9.4 Style of writing

▶ **Write as you would speak in comfortable, serious conversation**

▶ **Use plain language.**

Keep the style clear, simple, and fresh. Letters must be clear to nonnative speakers of English. Imagine that you are across the table from or on the telephone with the person to whom you are writing. Write in the way you would speak in these situations, but without colloquialisms or contractions ("don't," "wouldn't," etc. See Section 17.3.1).

▶ **Make the letter sound courteous and personal.**

Even if the letter to you seems rude or hostile, it is essential to thank the writer courteously for it in the opening paragraph of your letter and maintain courteous phrasing throughout.

▶ **Put yourself in the reader's mind, and work out how he or she would react to your language.**

It is possible to write something innocently that could be interpreted quite differently by the reader. For this reason, stand away from your personal involvement in what you have written and try to interpret it in the way the reader may see it.

▶ **Avoid the old-fashioned, pompous phrases associated with classic formal letter writing. Express the idea in plain English.**

Do not use phrases such as:	*Use these instead:*
As per	In accordance with
Attached hereto or herewith	I am *or* We are attaching *or* Attached is …
Enclosed hereto or herewith	I am *or* We are enclosing *or* Enclosed is …
Pursuant to your request	Following your request …
We are in receipt of your letter	Thank you for your letter
We wish to advise	We are pleased to tell you that/let you know that …
You are hereby advised	This letter is to let you know that …
Please contact the writer	Please contact me.

▶ **Avoid clichés, jargon, and organization-speak.**

▶ **Make sure the spelling and grammar are correct.**

Spell-check at the very end of writing. However, proofread the letter thoroughly after it has been spell-checked, since the spell-checker can pass words that you did not intend. If you know that your grammar may be faulty, give the letter to someone else to check.

9.5   Sample letters to illustrate the principles

The two sample letters shown here illustrate some of the principles of formal letter writing. They are written in response to a letter and a phone call from an irritated member of the community to a local government organization.

The two letters contain the same information. The first is impersonal, uses outdated phrasing, and is abrupt and uncaring. The second is friendly and personal while remaining appropriate to the formal situation, and it would be easily understood by a nonnative speaker of English.

Poor example

Use of impersonal greeting "Dear Madam."	Dear Madam,
	Litter infringement notice
Pompous, outdated wording: "As per;" "re." Personnel unnamed.	As per previous correspondence re litter infringement policies, I have followed up your query with the relevant personnel.
Pompous, abrupt, curt wording.	Your attention is drawn to the letter you were sent on October 8. A copy is attached. Regarding your phone call, when our staff attempted to return it on several occasions, it appears that the phone was unanswered.
Old-fashioned wording: "I wish to advise," "pursuant to…".	I wish to advise you that pursuant to the *xxx* Act 1991, you remain in infringement of Sections 92 and 115 despite your protests.
Abrupt instruction to contact the impersonal "writer."	If you have any further queries, please contact the writer.
	Yours sincerely,
Lack of first name: impersonal, no gender.	J. White Refuse and Recycling Officer

Better example

Greeted by name.	Dear Mrs. Smith,
	Litter infringement notice
Use of "Following" rather than "Further to." Reference to another person by first and family name—sounds personal.	Following your letter of October 4, 2006, and your phone call, I have followed up your query with Jim Brown, who administers the litter infringements.
Main point graciously stated, followed by apology and explanation.	We feel that the litter infringement notice should remain in place since you are in infringement of Sections 92 and 115 of the *xxx* Act of 1991. I am sorry that it has upset you. However, I hope that you will appreciate that this is an effective way of stopping people from misusing the public litter bins, which is a considerable problem in the city.
Pleasant reference to previous communication.	Attached is a copy of the letter that Mr. Brown sent you on October 6. Regarding your phone call, he did try returning it on several occasions but there was no answer.
Pleasant invitation to contact the named writer, using "me" instead of the impersonal "the writer."	If you have any further queries, please do not hesitate to contact me.
	Yours sincerely,

Signed with first and family names: personal touch.	John White Refuse and Recycling Officer

9.6   Letters that accompany a document

9.6.1   Cover Letter

Any letter that is sent together with any document.

Purpose
- To provide the recipient with a specific context within which to place the document.
- To give the sender a permanent record of having sent the material.
- To show willingness to provide further information.

9.6.2   Letter of Transmittal *or* Letter of Submission

A letter that accompanies formal documents such as reports or proposals.

Purpose
- To identify the topic of the document and its scope or extent.
- To highlight particular features of the document.
- To give an overview outlining the structure and contents of the primary document.
- To identify the person who authorized the document.
- To provide formal authorization of the date of submission of the document.
- To call for a decision or other follow-up action.
- To emphasize any particular points you may want to make.
- To show willingness to provide further information.
- To provide names of people to contact.

Additional purposes for a proposal
- To tell the client which companies have prepared the proposal.
- To point out the main selling points of your proposal.

Structure
A Letter of Transmittal should be *brief:* no more than one page long.

Some organizations include in the Letter of Transmittal a summary of the main points of the document. However, it is more effective to treat the Letter of Transmittal as a formal accompanying letter and provide an effective Executive Summary within the document (see Chapters 3 and 6).

▶ *First paragraph*
- Thank the client for the opportunity to submit a tender for the project.

- Describe what is being sent and the purpose of sending it.

▶ *Middle section*
  - Summarize the key selling points of the proposal (as decided by you or as called for in the Terms of Reference).
  - Give the reasons that the organization is best suited to carry out this work because of previous relationships, experience, expertise, resources, and other differentiating points.
  - If necessary, outline any critical aspects of the project.
  - If appropriate, give details of any alternative proposal that may be submitted in addition to a conforming tender. These will have been discussed with the client before the submission of the proposal to ensure that they are acceptable to the client.

▶ *Final paragraph*
  Establish goodwill by thanking the recipient, provide contact names and details, and show willingness to provide further information.
  Make sure that the letter also includes:
  - The project title;
  - The name of the organization or consortium submitting the proposal;
  - The client's project reference number;
  - The names of the organizations within the consortium;
  - The lead consultant's name and contact details; and
  - Personal signature of a member of the management personnel.

Format

▶ Use the standard format for a formal letter (see previously). Use the same font as is used in the main document(s).

Checklist for formal letters

The parts of a letter
- ☐ Is the letter left justified, except for the subject heading?
- ☐ Is the date formatted correctly, e.g., October 8, 2001?
- ☐ Is the name and address of the person to whom you are writing left justified, no commas at end of each line?
- ☐ Does the salutation properly address the person: Dear Sir, Dear Madam, Dear Sir/Madam: Dear Mr. … /Mrs. … /Ms. … /Dr. … /Prof. …
- ☐ Does the subject heading describe the main point of the letter?
- ☐ Is the subject heading centered and boldfaced?
- ☐ Did you style the closing correctly ("Yours sincerely," or "Yours faithfully")?
- ☐ Did you include your signature?

☐ Did you include your typed name, including your first name, not just your initials and surname?

☐ Did you include your position?

☐ Did you include the abbreviation "Encl." (if you are enclosing something)?

Font, spacing, arrangement

☐ Did you use a simple font, 10 or 12 points?

☐ Did you use single line spacing?

☐ Did you insert one blank line between paragraphs?

☐ Did you use a pleasing arrangement on the page?

Structure of the information

☐ Is the first paragraph courteous? If appropriate, does it acknowledge the letter/phone call you are responding to?

☐ Is your main point at the beginning of the letter?

☐ Is your supporting information after the main point?

☐ Does your final paragraph make a courteous finish?

Style of writing

☐ Did you use simple language?

☐ Did you avoid old-fashioned standard phrasing, clichés, jargon, and organization-speak

☐ Is the letter easy to understand for someone without your level of knowledge?

☐ Did you ensure that your phrasing could not be misinterpreted?

10

Short Workplace Documents

E-mails, Faxes, Memoranda, Agendas, and Minutes

Much formal professional information is communicated in the workplace by short documents that are produced in-house. E-mail and fax have mostly superseded memoranda, which were formerly the method of choice. Agenda and minutes are produced for formal meetings that require written records; they are also an aid to focusing the meeting's structure and discussion.

There are long-standing conventions governing memoranda, agenda, and minutes. In contrast, the e-mail system has novel features as a means of communication, in particular the immediacy of both receiving and sending, and its digital as opposed to hard copy storage. This has prompted many organizations to set policies of etiquette for its use.

Faxes differ from e-mails in that, although the transmission is immediate, they are hard copy and therefore more permanent.

Structure of the chapter

10.1   E-mails to communicate matters of work

Because of the immediacy and sometimes informal nature of e-mail, many organizations have policies for its use in communicating matters of work. Here are guidelines that are commonly in use.

▶ **Style of writing.**
   • Take as much care writing an e-mail as you would writing a letter. Be careful what you say and how you say it.
   • Do not use the pop conventions of the e-mail culture. Lower case letters at the start of sentences—*i* instead of *I*, *u* instead of *you*—will make a poor impression.
   • For people you know, it may be appropriate to start the message with their name followed by a colon.
   • For someone you do not know or are on formal terms with, start with the conventional Dear Mr./Mrs./Ms./Dr. (see Chapter 9). Finish with the corresponding closing.
   • Structure the content of your message in the same way as you would a letter. Do not do a brain-dump and let your text become unstructured.

▶ **Confidentiality: Assume that mail traveling via the Internet is not confidential.**
Never put in an e-mail message something that you would not want other people to read.

▶ **Permanence: Do not regard your files, either sent or received, as in safekeeping.**
Networks are not fail-safe. Print out hard copies of anything important.

▶ **Commercial sensitivity.**
No commercially sensitive material should be sent by e-mail.

▶ **Contractual material.**
Avoid using e-mail for contractual material unless it is followed by hard copies.

▶ **Attachments.**
   • When sending or receiving attachments, scan them for viruses.
   • Check the size of file attachments before you send them. If they are large, zip the file first; this avoids transmission decoding problems.
   • Any files sent via e-mail must have the permission of the author.

▶ **Unnecessary messages.**
Do not send unnecessary messages, particularly when forwarding material to large groups. The minor effort involved in doing it is far outweighed by the irritation it may cause.

▶ **Content of the autosignature.**
Make sure that your autosignature contains your name, address of your institution, and telephone and fax numbers. You may also want to include such

items as the URL of a personal web site. If your system does not carry an auto-signature function, then make up a template and use that for each message.

10.2   Faxes

As with e-mails, take as much care as you would when writing a letter (see Chapter 9).

Guidelines for writing it

►   If possible, use a fax template for the cover document. This will lay out all the necessary material such as recipient's name, institution, fax number, and so on.

►   If you are faxing to someone you do not know or with whom you are on formal terms:

- If possible, do not handwrite it.
- Use the conventions for starting and finishing letters (see Section 9.1.2).
- Structure the content of the fax as you would a letter (see Section 9.3).

10.3   Memoranda

Purpose

A memo (short for *memorandum*) is a very short document, usually up to a page long, that transmits information internally. In effect, it is an adaptation of a business letter. The memo is now being rapidly superseded by e-mail for short messages that are less formal than a letter. The most common types of memos are information memos and recommendation memos.

Guidelines for writing it

►   *The memo heading:* The distinctive element of a memo is the formatting of the preliminaries. They are quite different from those of a letter. A memo is headed by the word MEMORANDUM, followed by the side headings:

To:
cc:
From:
Date:

Then either the side heading "Subject:" or (in a longer memo) a centered title.

The use of these headings is shown in the following sample structures.

10.3.1   Memo structure for two types: a very short memo and one of about a page long

A very short memo (about half a page long)

Example

MEMORANDUM

All headings	**To:**	*(Name and title of the person to whom you are writing)*
justified to the	**cc:**	*(Names and titles of other people to be sent copies of*
left-hand margin.		*your memo, if required)*
	From:	*(Your name)*
	Date:	*(In the style October 8, 2006, not 10/8/06)*
Title, see Section	**Subject:**	*(A clear, informative title, containing the main message*
5.2.1, Chapter 5.		*of your memo)*

In separate   **Purpose of the memo.**
paragraphs.
Your main point or conclusion(s): Note that the supporting data are placed after the conclusion(s) or main point. Do not lead up to them and place them at the end.

Facts/data to support the main point or conclusion(s): See Section 9.3.

Recommendation (if necessary).

Your signature: *Note:* No salutation or closing, i.e., do not use "Yours sincerely/faithfully." Compare details of the individual elements, Section 9.1.2.

A longer memo (about a page long)

► **Begin with a short summary.**
Even though it is a very short document, your readers will appreciate having a brief summary. This will orient them so that they can better assess the information in the main part of the memo. See Chapters 2 and 6.

► **Organize the topics of the main body of the memo in order of importance.**
Put the key statement first, details afterward.

► **Use side headings and white space.**
Do not give your readers a page of unbroken text; it will look boring and daunting. Even in a one-page memo, use appropriate side headings and format it for free space.

The following is an example of the structure for a longer memo:

Example: A recommendation memo

Purpose
To come to a conclusion and make recommendations concerning an issue that you have investigated.

Suggested headings and structure

Example

MEMORANDUM

All headings	**To:**	*(Name and title of the person to whom you are writing)*
justified to the	**cc:**	*(Names and titles of other people to be sent copies of*
left-hand margin.		*your memo, if required)*
	From:	*(Your name)*
	Date:	*(In the style October 8, 2006, not 10/8/06)*

See Section 5.2.1,
Chapter 5.

Title

An informative title, centered, boldfaced

See Chapter 6,
Summarizing.

Summary: Very briefly state:

- The purpose of the memo
- Your main conclusion
- Your main recommendation

See Section 5.4.2, **Purpose:** A brief statement of why you are sending the memo
Chapter 5.

See Section 5.2.3, **Background:** A brief description of the background to the work
Chapter 5.

Further appropriate heading(s)

See Sections 5.2.4
and 5.2.5, **Conclusions and Recommendations**
Chapter 5.

Your signature: Note that there is no salutation or closing
(i.e., do not use "Yours sincerely/faithfully," etc.)

Compare the conventions: the elements of a formal letter, Section 9.1.

10.4 Agenda and minutes of a meeting

The agenda of a meeting tells the participants what topics will be discussed at the meeting. Minutes record what occurred: the decisions taken, the form of the discussion, actions to be taken, and so on.

10.4.1 Agenda

An agenda of a meeting is a listing in sequence of the topics to be discussed. The topics are not described in any detail; the agenda is a simple list of headings, together with the names of the people who will be introducing or leading discussion of the individual items.

Agenda are circulated before the meeting to allow participants to prepare their contributions to the discussion.

Example

Bright Consultants Training Committee
December meeting
9:30 a.m., Friday, December 10, 2006, Meeting Room 2.1

AGENDA

1. Minutes of the previous meeting: Friday, November 12, 2006.
2. Matters arising from the minutes.
3. Report by A. White on the 2006 training program.
4. Report by B. Brown on specific requirements for further programs.
5. Discussion on program for 2007.
6. Review of deadlines.
7. Agenda for next meeting.

10.4.2   Minutes

Minutes are an essential part of an organization's documentation. They there-fore must be accurate and carefully written.

Guidelines for writing it

- ▶ Overall, clear and concise points are preferable to a blow-by-blow account of a long, unstructured meeting.
- ▶ State the name of the group that is meeting.
- ▶ Record the time and place of the meeting.
- ▶ List the names of all the people at the meeting at the beginning of the minutes, together with any apologies for nonattendance and the names of those who are absent.
- ▶ Record whether the minutes of the last meeting were approved or approved with amendments.
- ▶ State whether there were any matters arising from the minutes of the last meeting.
- ▶ In the order of items on the agenda, make sure that each major point that was discussed is recorded.
- ▶ Note the main contributors to each topic, the decisions taken, and the individuals responsible for each necessary action.
- ▶ Record any formal motions: the names of the proposer and seconder for each motion, and the result of the vote on that motion.
- ▶ State the time that the meeting was formally concluded.
- ▶ Record the time, date, and place of the next meeting.

The minutes of a meeting are usually distributed to participants before the next meeting. This allows the participants to check whether the minutes are a true and accurate record of the events. A vote to formally pass the minutes is usually required in the next meeting.

If the minutes are confidential, each page should be marked with a message such as CONFIDENTIAL: DO NOT DISTRIBUTE.

Here is an example of a suggested layout of the minutes of the committee meeting described in the preceding agenda.

Example

Bright Consultants Training Committee
Minutes: December meeting
9:30 a.m., Friday, December 10, 2006, Meeting Room 2.1

Present: A. White (chair); J. Black, J. Green, S. Brown.
Apologies: F. Smith. Absent: G. Jones.

1. The minutes of the meeting on Friday, November 12, 2006, were approved with no discussion.
2. There were no matters arising from the previous minutes.
3. Ann White reported on the 2006 training program. She noted that it was successful in overall terms, and had received good feedback from the participants. However, the word-processing and spreadsheet programs were at too advanced a level for some of the participants. She suggested that simpler programs be instigated for beginners, while retaining the present structure of the various courses. There was general agreement. It was suggested that CompuTrain be approached for details of its beginner programs.
   ACTION: J. Black to discuss with CompuTrain.
4. Ben Brown presented his ideas on specific requirements for further programs. Jack Green felt there was a need for more rigorous training in engineering report writing, and that ...
   ACTION: A. White to explore possibilities of advanced courses in engineering report writing.

[etc.]

The meeting closed at 10:30 a.m.

Next meeting: 9:30 a.m., Friday, February 11, 2007, Meeting Room 2.1

Checklist for short workplace documents

E-mails

☐ Did you use the conventions for the salutation and the closing of a letter?

☐ Is the content of the e-mail structured as you would structure a letter?

☐ Did you avoid sending commercially sensitive or contractual material by e-mail?

☐ Did you make hard copies of important e-mails, both sent and received?

☐ Do you scan attachments for viruses?

☐ Do you zip large files if they are to be sent as attachments?

☐ Do you avoid forwarding unnecessary messages?

☐ Does the autosignature contain your name, address, and telephone and fax numbers?

Faxes
- ☐ Have you, if possible, used a fax template for the cover document?
- ☐ Does the fax use the conventions for starting and finishing a letter?
- ☐ Is the fax structured like a letter?

Memos
- ☐ Have you used the conventions for the preliminaries of a memo?
- ☐ Is the memo structured so that the key statement (your main point) is at the beginning of the memo, and not at the end?
- ☐ For a longer memo: Is it structured under suitable headings, with use of white space?
- ☐ To close the memo, have you avoided using *Yours sincerely* or similar wording?

Agenda
- ☐ Is it a simple listing of the topics, with the names of the people who will be introducing or leading discussion of the individual items?
- ☐ Does the agenda state the time and place of the meeting?
- ☐ Will the agenda be circulated to all the meeting's participants well before the meeting?

Minutes
- ☐ Do they state the name of the committee or group and list those who were present, together with any apologies or absences?
- ☐ Do they state that the minutes of the previous meeting have been approved?
- ☐ Do they itemize the matters arising from the previous minutes, if any?
- ☐ Is each major topic discussed at the meeting covered (the principal contributors, the actions that need to be taken, the people charged with those actions)?
- ☐ Are the formal motions (if any) recorded (their proposer and seconder, the result of the vote)?
- ☐ Are the time of conclusion of the meeting and the date, time, and place of the next meeting stated?
- ☐ If the minutes are confidential, is each page marked with a warning?

11

Publicity Material

Brochures and Press Releases

Brochures and press releases act as advertisements for an organization. Brochures are written and designed for rapid assimilation of essential information. Technical press releases usually announce the development of a new product or a significant change in the structure or personnel of an organization. They may be aimed at a wide range of readership, but as a first step they need to be written so that journalists and editors will assess them for inclusion in their trade journals, magazines, or newspapers.

Structure of the chapter

11.1   Writing a brochure

This section covers only the basic aspects of writing a brochure. It does not cover project management, budgeting, scheduling, design, or use of the brochure.

Purpose
- To be a basic marketing tool.
- To be portable and fileable.
- To establish your organization's credentials and legitimacy.
- To convey the essential information the client needs, while projecting an image that sets your organization apart.

What to keep in mind

The conventional wisdom is that readers will spend only a minute or two on a brochure. They will glance at the photos and read the headings, subheadings, and anything that is highlighted. They may then scan the list of clients or projects and possibly read isolated paragraphs, especially the first and the last. Make the best of these elements. Remember, though, that others will read the whole brochure intensively.

Guidelines for writing it

▶ *Define the message and the audience.* You will then be able to define the format and style. Basic questions to help define message and audience are
   - What is the company's objective in producing this brochure?
   - What is the impression you want to make?
   - Who are your potential clients?
   - What are your clients' concerns and problems?
   - What are your company's strengths?
   - What are your company's weaknesses and perceived weaknesses?
   - Who are your competitors? How are you different?
   - What are the most important projects/services/geographic areas to be covered?
   - How do you get projects?
   - How will the brochure be used?
   - How many copies will you need?
   - What is your budget?

▶ *Choose a format,* e.g., single-sided, double-sided, large sheet folded into two or more folds, presentation folder.

▶ *Gather ideas on content and style.* Examples of basic content are
   - A description of the business group/market sector
   - How you differ from your competitors
   - The company's approach to clients
   - Services available
   - Types of projects undertaken
   - Examples of projects and previous clients
   - A list of the company's offices

   If appropriate:
   - Client testimonials. They add credibility.
   - Message from one or more key people on the staff or a description of their special expertise.

It is advisable to engage a specialist writer for the copy. Their experience will ensure that the content, style, and tone are appropriate.

▶ *Avoid the following*
   - Information that will quickly date your brochure, e.g., phrases such as "to be completed in 2006" and "is now under construction."
   - Photos of staff identified by name. They could leave the organization or visibly age during the brochure's life and make it look dated.

- Poor-quality photos. Every photo should be excellent. If not, it is better to have none at all.
- A brochure that looks crammed. Avoid the temptation to fill the page; white space is absolutely necessary around the photos (to create a frame) and around the text (to make the information more readily assessed).
- Typographical errors. Your brochure should reflect quality and attention to detail. See Chapter 17.

11.2   Writing for the media

Publicity differs from other types of self-promotion such as advertising because it is assessed and endorsed by a neutral third party: the media. The media has the final decision about what gets published.

Journalism is a difficult form of writing because it involves writing simply. Complex issues have to be put into everyday terms. It also has to conform to space restrictions, be produced under tight time schedules, and follow a certain format.

11.2.1   Benefits of getting published

- Gives public recognition of an organization's activities.
- Builds the organization's image.
- Helps promote the client organization and sell its project.
- Gives an objective endorsement—and therefore credibility—by a neutral third party, and is therefore free of obvious self-promotion.
- Provides reprints for promotional use.

11.2.2   Objectives of getting published

- To expand existing markets or enter new ones.
- To communicate the organization's capabilities.
- To attract better employees.

11.2.3   Selecting the publication

Possible types of publications are
- General press and media
- Regional and local media
- Business publications
- Trade and industry magazines specific to market sectors
- Professional and trade association journals
- International publications

You can also do the following:
> ► Ask clients and specific targeted audiences what they read and what publications influence them.
> ► Note the publications in their reception areas or bookshelves.
> ► Ask a publication to send you its media kit.
> ► Consult editorial assistants and sales representatives.

11.2.4   Types of media releases

A story can be given either a "feature" or a "news" angle.
- The feature is usually a reasonably lengthy article that includes a good deal of background and opinion relevant to the subject it covers. It provides for considerable flexibility of style and for personal viewpoints. The aim is to attract the reader in the opening paragraph and then involve and hold the reader's interest as the story unfolds.
- The news story is factual and should include answers to the six W and H questions—What, Where, When, Who, Why, and How.

11.2.5   Your aim when writing a media release

- To give readers of the mass media the main points, in as few words as possible, as simply as possible, in a style that is interesting and attractive.

11.2.6   Guidelines for writing for the mass media

> ► Determine the main point of your article.
> ► State the most important issues first. Write the material so that no important factors will be lost if the release is cut from the bottom by the newspaper copy editor. This is referred to in the media as "writing from the top" or "inverted pyramid writing." This means leading with the most important and interesting points, because media consumers seldom read all of a story—they read the first part and then scan the rest. See Section 11.2.8.
> ► Be accurate, brief, concise, and factual. This establishes credibility with editors.
> ► Emphasize the main point, the project, the challenges and solutions, and the client.

Avoid the following:
> ► Avoid introverted writing—writing in relation to one's own needs and attitudes and intellectual capacity. Most nonprofessional writers write for themselves or their peers. Write for someone unassociated with the subject reading what you are writing.
> ► Avoid putting your company's name in the first paragraph. Many organizations believe this gets their name in front of the media. This

can result in a high failure rate. Example: "The XYZ organization today announced that. . .". This structure can become tedious to news editors and they may not bother to read on to where a major announcement is made.

► Avoid praising the organization or its people.
► Avoid long words if short ones will do; avoid clichés, superlatives, glossy adjectives, and jargon.
► Avoid unnecessary or redundant words. Common examples: "almost unique" ("unique" means one of a kind; there is no such thing as almost unique, rather unique, or really unique); "new initiatives."

Be aware of the following points:

► Approvals might be needed within your organization and from the client.
► If you need to gain approval from the client, submit copies of all material (text and visuals). Explain the intent of the submission and where it will be published. Receive approval *in writing*.
► Photography and visual images are critical. Choose a professional photographer according to expertise and experience with similar types of assignments. Ask targeted publications for recommendations.
► Know photographer's copyrights and fees.
► Do not expect photos and slides to be returned. For the initial inquiry stage, submit duplicates.
► Do not expect to see the text before publication.
► There is often a long lead time before publication. Many national publications close an issue months before press date.

11.2.7 Writing a news item

► **Determine what is news.**

Before starting work on any release, ask yourself, "What's the news value of this story? What's going to interest most people?" News is about *now*. A decision taken at last night's meeting is news to this morning's newspaper. For a weekly newspaper, it is news in this week's edition. By next week, it is no longer news.

Remember that the discard rate for unsolicited press releases at most newspapers is very high.

► **Ask the six W and H questions.**

The rule of thumb that cub reporters are taught about writing a news story is simple: stories must answer the What, Where, When, Who, Why, and How set of questions. The order is not important, but that is the basic information the editor will be looking for. A story about something about to occur will answer the same questions in the future tense.

▶ **Keep it short.**

Try to keep it to one page, never more than two. Use simple words, short sentences, and short paragraphs. Put in the relevant facts but do not bore the editors with all the background and details. If the editors want to know more, they will get in touch with you.

▶ **Keep the language simple.**

The article must be clear to the average person. It is said, somewhat cynically, that the popular press caters to 10-year-olds. This does not mean that writers must write down to people. Anyone can recite complex technical terms and definitions; the person who really understands a subject is the one who can explain it simply.

Do not use jargon, flowery adjectives, clichés, or unnecessarily long words. *Mark Twain:* "I never use the word metropolis when I can get paid the same amount for the word city."

▶ **Put a source on the statement.**

It is important to correctly attribute what is said. Instead of saying that the organization believes such and such, put the words into the mouth of a person: your CEO or the Manager of Environmental Planning. There must also be a clear distinction between fact and opinion. What an organization thinks is fact may be opinion to other people.

Example
If government subsidies are reduced, employment in the industry will fall drastically.

Why this is not acceptable
Who says so? This is not a fact that can stand alone as a bald statement. If a company believes this to be the case, then the release should state who said it.

Example
"If government subsidies are reduced, employment in the industry will fall drastically," the Vice President of the XYZ Company, Mr. John Brown, said.

Why this is acceptable
Attributed to a specific person in the organization.

11.2.8 Writing the intro or lead paragraph of a news item

Purpose
To first attract the news editor's attention, then the readers' attention.

Guidelines for writing it
▶ Place the news first. The first paragraph must contain the critical element of a newspaper story. This is the selling point of the story.
▶ Make it no longer than 25 words.
▶ Use the active voice of the verb rather than the passive voice (see Chapter 17).

Example

> Built for just $10,000, a mini-jet airplane was given its maiden flight over Newtown air base on Saturday by a 21-year-old apprentice engineer *(name)*, who constructed it in his backyard.

This is not a media introduction as it stands, because:
- The cost of the plane is not the main point. The interesting part of the statement is buried at the end: the fact that the plane was built by a 21-year-old in his backyard.
- It is more than 25 words.

Example

> *A journalist would produce an intro paragraph something like this:*

> A 21-year-old apprentice engineer test flew a homemade mini-jet airplane at Newtown on Saturday.

> *Secondary information will then follow:*

> The jet, which he built in his backyard for just $10,000, was tested over Newtown air base by its maker and designer, (name).

This is a media introduction as it stands, because:
- It takes the most important and interesting point and uses it first.
- It is only 19 words long.
- It uses simple terms (changes *constructed in his back yard* to *homemade*).
- It answers the six questions: what, where, when, why, how, and who.
- The original statement was in the passive voice: was given its maiden flight … by … .
- The remaining details are then ranked in successive order of interest.

11.2.9   Style of writing: keep it simple

Articles for the mass media are very different in style from other types of formal, work-related writing. Make sure that you do the following:
- Make the story easy to read.
- Be accurate. Double-check the spelling of names (give first names and middle initials), addresses, dates and days, and the exact title of a speaker, report, pending legislation, and so on. Also include honorifics: Ms., Mr., Dr., and so on.
- Keep the paragraphs short. Usually there should be one main point to a paragraph. The introduction or lead paragraph should be no more than 25 words. The succeeding paragraphs should be kept to 15 to 40 words.
- Use short sentences, with no more than one thought in a sentence. At most, include two ideas in a sentence if they are brief and related.
- If acronyms are used, spell out every term when it is first used and place the abbreviation in parentheses immediately following—Environmental Sustainability Index (ESI). Subsequently, the abbreviation alone can be used. For specialized media, commonly used technical abbreviations can be used.

- ▶ Use verbal abbreviations only in reported speech.
- ▶ Write all numbers from one to ten as words, numbers from 11 upward as figures.
- ▶ Journalists are constantly obsessed with deadlines. To be news, releases should carry words like "today," "this afternoon," "tomorrow," and so on.

11.2.10 Format for media releases

- ▶ Type and double-space or at least one-and-a-half space text, leaving wide margins on both sides and at the top of the page for editor's notes.
- ▶ Use single-sidedpages. Never use double-sided pages.
- ▶ Date clearly.
- ▶ Include name, address, and contact phone numbers of the writer for further information. Also include this information for anyone who is reported in the release.
- ▶ Staple pages together. Paper clips can become dislodged.
- ▶ Clearly distinguish your organization. Use letterhead if possible.
- ▶ Include directive information. Include the words "News Release," name of person to whom it is addressed (newsroom and reporter if known), your organization's address, and the subject of the release.
- ▶ List contacts. Give at least two names with home and business phone numbers of people in your group who can give further information and can be interviewed if the paper requests it.
- ▶ State at the bottom of each page "More follows," "More to come," or a numbering sequence such as 1 of 2 or 1/2 to show that more of the story follows and its full length. Mark the final page "Ends" or indicate the end by a solid line.
- ▶ Include a brief cover letter. Highlight the main points of interest to the specific publication.
- ▶ Do not worry about packaging. Bulky folders are often a nuisance to an editor.

11.2.11 Illustrations

If possible, include photos in releases to newspapers and magazines. They must be of good quality—clear delineation and good contrast—since the reproduction processes can blur pictures somewhat. Often you can let the paper know that you have a possible photo and they will send along their photographer. Remember to give them enough notice (at least 1 day and preferably several days). If possible, give editors a selection of photographs from which to choose.

11.2.12 Deadlines and embargoes

A deadline is the latest possible time at which the media can accept material. To deal with this, you need to be aware of the peak times for news preparation. If material arrives just before the deadline, the newspaper may be full. Morning newspaper staff work in the afternoon. Afternoon newspaper staff work in the early morning.

The essence of writing for deadlines is quick reaction and sometimes preparation of information for events before they happen.

The embargo system is a mutual agreement with the media: Material is made available in advance on the understanding that it will not appear until, on, or after a prescribed release time. For instance, if a vice president is to make a statement at lunchtime on Tuesday and the newspaper's deadline is 9 a.m. Tuesday, the material will be released to the newspaper on Monday or earlier on an embargo basis.

11.2.13 Additional strategies if your organization is small or unknown

▶ Look to trade and industry magazines. They are one of the best sources for reaching a market audience. Many of them welcome good projects because they are often overlooked in favor of national magazines.

▶ Write byline articles that establish the organization's expertise. Trade magazines are good for byline material because they often have a very small editorial staff. When writing a byline or "how-to" piece, keep the focus and subject strictly educational in nature and use projects to illustrate the points.

▶ Enlist the aid of industry vendors and be visible in their publications.

▶ Submit to award programs sponsored by magazines.

11.2.14 If your story is not used, do not be discouraged

It is estimated that only 10% of incoming news releases are used by a medium, and this includes the slick ones produced by expensive public relations firms.

Sometimes it has nothing to do with the quality of your release, but rather the unfortunate timing of a large event that crowds out other material. It becomes easier once the newsroom gets to know you and realizes that you are seriously seeking to share news with them.

12

A Journal or Conference Paper

A paper for a learned journal or conference describes some form of investigation written in the specific form demanded by the editor as given in the *Instructions to Authors*. The paper will usually follow the traditional format for research papers, at least in the basic skeleton of section headings. Research papers require full citations of others people's work (see Chapter 14).

This chapter gives information on the process of publishing a paper and the traditional structure of a paper, together with guidelines for writing the various sections.

Structure of the chapter

12.1 The process of publishing a journal paper
12.2 The structure of a journal or conference paper
12.3 Requirements for the sections of a journal or conference paper
Checklist for journal and conference paper

12.1   The process of publishing a journal paper

The length of time between submitting the final version of the paper to a journal and its appearance in the journal can be from a few months to more than a year. This section describes some of the processes involved in this often lengthy and sometimes frustrating procedure.

12.1.1   Choosing a journal

Factors that need to be considered:

▶ **The level of prestige of the journal.**

Those of greater prestige will have a higher standard and a greater rejection rate than those lower on the scale. Acceptance in a high-prestige journal will increase your professional standing.

▶ **The lead time to publication.**

Some journals have extensive delays before publishing. This may be due to a slow reviewing process, editorial delays, and the difficulties of production schedules. You can determine potential delays by asking other people who have published in the journal or by contacting the editor.

▶ **Whether the journal has page charges.**

Some journals publish papers with no charge to the author. Others charge quite a high rate per page, which may mean a substantial cost.

▶ **Costs of reprints.**

Some journals supply a given number of reprints free of charge; others charge for them. You will need to decide how important reprints are to you.

12.1.2 Formatting the manuscript

Journals have reasonably straightforward instructions about formatting and content. In some journals, they are available in each issue under headings such as *Instructions to Authors* or *Information for Contributors*. In some cases, the journal asks you to request a booklet from the editorial office. For many journals, Instructions to Authors are available on the Internet.

The instructions should be meticulously followed, even if you do not like the result.

Typical information in the Instructions to Authors includes the format in which the manuscript must be typed (font, type, size, margins, heading style, etc.); the style of the illustrations; the number of copies required; the requirements for the digital file; and the mailing instructions.

12.1.3 Submitting the manuscript

The stated number of copies of the typed manuscript, the digital file, and a letter of transmittal are usually needed.

12.1.4 The letter of transmittal

This is an accompanying letter sent together with the paper to the editor of the journal. It should be very simple, stating only that you would be grateful if the enclosed paper could be considered for publication in the journal. Address it to the editor; the most recent issue of the journal will give the name and address (see Section 9.6.2).

12.1.5   Mailing the manuscript

The Instructions to Authors will tell you the exact number of copies of the manuscript needed and the requirements for the digital file. Usually at least three hard copies are needed for reviewing purposes.

Since the package will be bulky, you need to make sure that the postal system will not damage it. Consider using a padded envelope (thick manuscripts can burst out of simple envelopes). In addition, backing cardboard should be used when photographs are included with the manuscript.

Send it by the most rapid system of mail. If you have not had an acknowledgment in two weeks, follow up with an e-mail, a phone call, or a letter.

12.1.6   The reviewing process

This depends on the type of journal. In general, the following happens:

The editor sends the manuscript to several reviewers, who may take some time in the reviewing process. The reviewers will then send their comments back to the editor, together with their recommendations about publication. Each reviewer may recommend any one of the following:

- That the manuscript is published with no alterations needed. This is rare. Most reviewers recommend some sort of amendment to the manuscript.
- That the manuscript is published subject to the reviewer's recommended alterations being made. These alterations can vary from small amendments of style to major changes of structure and/or content.
- That, if the changes needed are major, the manuscript will be once again reviewed before a final decision. If the required amendments are only minor, it may be necessary only for the editor to check to see that they have been made.
- That the manuscript is not suitable for publication.

On the basis of all the reviewers' decisions, the editor then chooses from the following possibilities: whether to accept your manuscript for publication subject to amendment, to subject it to a further review process after major alteration, or to decline it.

You will be told the decision in the editor's letter and also will be given copies of each of the reviewers' comments. The reviewers usually remain anonymous. Alternatively, the editor may send only a summary of the reviewers' comments.

12.1.7   How to deal with reviewers' comments and amend the paper

▶ **General advice**

All of the comments should be read with great care. Most reviewers' evaluations are usually helpful, and you can take advantage of a fresh viewpoint on both your writing and your work. The final version of the paper can be considerably strengthened by using the comments of a good reviewer.

▶ How to deal with comments that you do not agree with

If the suggested changes seem unnecessary or, with good reason, unacceptable to you, then the editor can be given a reasoned argument as to why you believe that a particular change need not be made. There are several things you need to address when evaluating a comment from a reviewer.

- *Does the fault lie with you or the reviewer?* Some reviewers' comments may show that they have misunderstood or misinterpreted your material. You then have to establish whether this is because you have not explained it well enough or the reviewer does not know what he or she is talking about. It is easy in the heat of the moment to assume the latter, but it needs careful reflection.
- *Is the reviewer possibly not an expert in the field?* The reviewers may not be the ultimate authorities on your topic. This may be no fault of the editor; it is extremely difficult to find appropriate reviewers for each of the hundreds of manuscripts that an editor has to deal with each year.
- *Are the remarks trivial?* In a few cases, you may have just cause to feel peeved. Some reviewers, if they are unable to make substantive comments, feel the need to justify their appointment by pointing out minor errors, for example, in the wording. Such comments can often reflect personal quirks and may not make for valid comments.
- *Are the comments dull, mechanical, or generalized?* A comment such as "Poorly organized" with no suggestions as to how it could be improved is a useless comment. A reviewer who is not lazy will make comments that are specific and will show a real interest in your topic.
- *Are all of a reviewer's comments negative?* This means either that the reviewer thinks your paper has no worth or that the reviewer is prejudiced or is trying to impress the editor.

Whatever your conclusions about the reviewers in terms of these questions, you cannot use words such as "lazy," "trivial," or "useless" in your rebuttal. If you decide not to abide by a reviewer's suggested amendment, you need to send the editor a calm, well-reasoned, and well-written defense that avoids pejorative terms. Your arguments should be contained in the formal cover letter when you resubmit your amended manuscript to the editor.

The editor will take note of your argument. If your facts are correct and your reasoning is sound, he or she will be able to use your argument as justification for reversing a negative decision.

If the reports from two assigned reviewers are contradictory, you can ask the editor for a third reviewer.

12.1.8 Resubmission of the manuscript in its amended version

Send the required numbers of copies of your amended manuscript and the file on disk to the editor. Your cover letter should indicate whether you have incorporated the reviewers' suggestions and should give reasoned arguments if you have rejected any.

12.1.9 If your paper has been accepted

The next stage is to receive the typeset version to proofread. This needs to be done meticulously; you will need to correct it using standard proofreading symbols (see Section 16.4).

Alternatively, the editorial staff of the journal may do the final proofreading.

12.1.10 If your paper has been rejected

Examine it very critically, alter it as you think necessary, and submit it to another journal.

12.2 The structure of a journal or conference paper

A journal paper often follows the classic Abstract, Introduction, Materials and Methods, Results, and Discussion (AIMRAD) pattern in its general format. Many papers will need these actual sections; others will need to contain the basic skeleton and follow the scheme in its general plan. It is essential to follow the journal's Instructions to Authors for structure and format.

12.3 Requirements for the sections of a journal or conference paper

12.3.1 Title

Purpose
To give the reader immediate access to the subject matter of the paper.

Guidelines for writing it
A title should be informative, not a generalized overview. Keywords are critical. You need to imagine how another person would look for this kind of information in a database. It would be a mistake to believe that a general title will suffice for a journal paper and that the list of keywords will indicate the specifics of the work. Many people choose papers to read from the titles in a List of References; an inadequate title may not be followed up.

A title can have various forms. A **hanging title** is when a colon or dash joins parts of the title. This is a useful way of avoiding a long, grammatically difficult title. Either the first or the second part of the title can be used to describe the overall area; the other part gives more specific material:

> Applications of Drag-Reducing Polymers in Sprinkler Irrigation Systems: Sprinkler Head Performance

Some journals allow the use of **questions** in the title.

> How Much Do Road Accidents Cost the National Economy?

Questions can also be used as the second part of a hanging title.

> Modeling Drainage Performance in Slums of Developing Countries: How Good Is Good Enough?

Some journal editors do not like **series titles**. If the various papers appear in different journals, there are problems with the timing of publication, with the result that papers can become out of sequence. However, they are still sometimes used, particularly when the individual papers of the series are published simultaneously in the same edition of the journal.

> Effects of Electroosmosis on Soil Temperature and Hydraulic Head. I: Field Observations

> Effects of Electroosmosis on Soil Temperature and Hydraulic Head. II: Numerical Simulation

Running titles (running heads) are the short titles required by journals for the tops of the pages. In contrast to the main title, running titles can use accepted abbreviations.

> *Full title*
> Dynamic Interpretation of Slug Tests in Highly Permeable Aquifers

> *Running title*
> Slug Test Interpretation

Abbreviations in the main title should be widely known in your discipline. Many journals have a list of the abbreviations they will accept.

Ensure that the title makes sense. The structure may be lost during the quest for the minimum number of words.

12.3.2 Authorship and affiliation

Purpose

To show the people who did the work presented in the paper, the institutions where it was done, and, if they have changed, the present addresses of the authors.

The journal's Instructions to Authors will define how to present the author/affiliation information.

12.3.3   Abstract

Purpose

To give the reader a miniaturized version of the paper: all of the key information—objective of the work, methods, results, and conclusions. For full details, see Section 6.9.

12.3.4   Keywords

This is a short list of words relevant to the work that will be used by electronic indexing and abstracting services. It is important to determine the keywords that a potential reader might use to search for information. The list should include both general and specific items.

12.3.5   Introduction

Purpose

- To clearly state the purpose of the study.
- To allow readers to understand the background of the study without requiring them to consult the literature themselves.
- To identify the authors who have worked or are working in this area and to describe their chief contributions.
- To point out the relationships between the various authors' works.
- To indicate correlations, contradictions, ambiguities, and gaps in the knowledge.
- To outline the approach you have taken with respect to the correlations, contradictions, ambiguities, and gaps.
- To provide a context for the later discussion of the results.

Guidelines for writing it

▶ *The first sentence* should provide an overall introduction, specific to the work described in the paper. Avoid making a banal statement of general knowledge or resorting to a trite statement of the obvious. For example:

Toxic waste is a very serious problem in the world today.

Even pompously dressing it up cannot disguise a banality:

The quantity of toxic waste currently generated in the world is a problem of the utmost seriousness.

The main purpose of the work should be clearly stated in the Introduction. Many readers point out that this is often missing or difficult to determine.

▶ Review the literature and show the relationships between the various areas of work. Show the background of the previous work in this area. Show the contributions of others, with correct reference citations of their work. The references cited should be carefully chosen to provide the most important background information.

▶ Show where there are correlations, contradictions, ambiguities, and gaps in the knowledge. Show the scope of the problem and how your work will address these issues.

▶ Make it simple and brief, but keep your reader adequately informed.

▶ Define the specialized terms used in the paper.

Structuring the Introduction

The Introduction tells a story—it should have a logical flow:

- *The beginning:* Briefly summarize relevant current knowledge, supporting your statements with references as necessary.
- *The middle:* Describe what is not known (or a problem with the known). Having summarized the established facts, move on to areas where there is less or no knowledge, or where the evidence is conflicting.
- *The end:* In the final paragraphs, clearly state the purpose of the work then briefly summarize your approach, if this is appropriate.

Every study sets out to solve a specific research problem. This should be stated explicitly in the final paragraph of the Introduction. Make sure it follows logically from the preceding sentences; these should have been structured so that the gaps or controversies in the knowledge are obvious. Use signaling words and phrases to introduce the purpose:

However, it is not known whether . . .
To answer this question (*such and such*) was investigated . . .
To clarify the role of X in Y, . . .
To determine whether . . .
To compare the properties of A and B . . .

Having stated the purpose of your work, it is sometimes appropriate to very briefly state how you did it.

Tense of the verb for the Introduction

See the Quick Reference Guide: Parts of Speech and Verb Forms (Part 7) for guidelines for using tense in technical documents and definitions and examples of the various tenses of the verb.

▶ Use a mixture of present and past tenses: the present tense to describe the established body of knowledge and existing situations, the past tense for people's findings.

Example to show use of present and past tenses in an Introduction

The resistance of buildings to wind pressure has been *(past tense)* the subject of considerable research. Normal design loads are *(present, established knowledge)* substantially lower than those that can occur in a severe windstorm. Furthermore, many common construction practices produce *(present, existing situation)* connections that are inadequate to resist loads in such severe windstorms. The development of retrofit options for improving the connection between building components has been *(past)* the subject of previous research. Results have been used *(past)* to develop the recommendations for improving the attachment of sheathing and for strengthening structural connections presented in this paper.

Adapted from Reinhold, T. A., Schiff, S. D., Rosowsky, D. V., and Sill, B. L. (2002). "Case for enhanced in-home protection from severe winds." *Journal of Architectural Engineering,* **8** (2), 60–68.

▶ Use the past tense for specific findings that you are going to dispute.

Example

Jones and Brown reported that airborne pollution was *(past)* not responsible for acid rain. However, other studies have not confirmed this finding.

Common mistakes

- The main point is not clearly obvious—the reason for doing the study is not clear.
- The literature has not been adequately reviewed. For example, the pivotal references may not have been cited; only a few references may have been cited for a thoroughly researched area of work; the correlations and contradictions may not have been pointed out.
- Excessive length and rambling, unspecific, unstructured, irrelevant material.
- Insufficient length with overly general material.
- Does not summarize the approach taken.
- Specialized terms are not defined.

12.3.6   Materials and Methods *or* Procedure

This section is often the easiest part of a document to write. Describing experimental methods is usually very straightforward. Therefore it is often the best place to start writing. There is no need to write a paper in sequence from beginning to end. Start with the section that is going to give you the fewest problems.

Purpose

- To describe your experimental procedures.
- To give enough detail for a competent worker to repeat your work.
- To describe your experimental design.
- To enable readers to judge the validity of your results in the context of the methods you used.

Guidelines for writing it

- ▶ Logically describe the series of experimental steps so that a competent worker in your field could repeat the whole procedure.
- ▶ Ask yourself whether you might be too familiar with the techniques. You might make the mistake of leaving out descriptions of procedures that are essential but which you take for granted. If you think this is the case, give your description to a colleague to read.
- ▶ Make sure that some of the results are not accidentally introduced. The Materials and Methods section and the Results section need to be very

strongly separated from each other in their contents. However, if you need the results of one experiment to justify using the subsequent methods, it should be acceptable to say so, briefly.

▶ Summarize established techniques; novel techniques or variations on an old technique should be fully described.

▶ Tables can also be used in the Materials and Methods; they do not belong only in the Results. A table is often the best way to describe a complex procedure.

▶ If you are following the conventional AIMRAD format, use subheadings. When possible, use subheadings that match those that will be used in the Results. The reader can then correlate a particular method with the related results.

▶ Most academic assessors and journal editors allow the occasional use of *We* in an active construction.

▶ When you need to cite a technique, cite the earliest reference in which this form of the technique was used.

If you have to refer to the literature to explain a technique, give enough information for the reader to get an outline of the technique.

Good
Specimens were concrete prisms with a deformed bar embedded at their center of the cross section and carbon fiber sheets externally bonded to their two side surfaces.

Poor
Specimens were prepared as previously described (Ueda et al., 2004)

Tense of the verb
See the Quick Reference Guide: Parts of Speech and Verb Forms (Part 7) for guidelines for using tense in technical documents and definitions and examples of the various tenses of the verb.

▶ For experimental work, use the past tense. You are describing work that you did.

Correct
Ten columns served as reference and were tested without any strengthening.

Incorrect
Ten columns serve as reference and are tested without any strengthening.

▶ For description of geographic or geologic features, use the present tense.

Example
All three paleosols show a greater degree of development than the surface soils. Better development is displayed in terms of greater clay accumulation, higher structural grade, harder consistency, and thicker profiles.

Common mistakes
• Not enough critical detail to enable someone unfamiliar with the method to repeat it. It probably happens because the techniques are too familiar.

- Too much unnecessary detail.
- Detailed text where an illustration would be more appropriate.
- Illogical description. This may happen when several procedures are described together.
- Being referred back to the literature with not enough summarized information to be able to understand the method.
- Introducing some of the results.

12.3.7 Results

The Results section is the core section of the document because it presents new knowledge. Data is provided in figures and tables that accompany the text.

Purpose

To present your results, but not to discuss them, giving readers enough data to draw their own conclusions about the meaning of your work.

Guidelines for writing it

► Highlight the most important aspects of the results in the text of the Results section. You need to guide the reader to decide what to look for in the tables and figures. A Results section should not be made up solely of illustrations; there must also be explanatory text linking them. For example, to show the behavior of various types of panels under load, it is not sufficient to present only a graph. The main features also need to be pointed out in the main text:

Example
The panels with angle-ply reinforcement behaved similarly to the central panels and showed no significant increase in load capacity (Figure 4).

► Make illustrations as self-explanatory as possible by means of good titles and captions. After reading only the title and Abstract, readers familiar with the topic will often turn next to the Results section. Moreover, studies of how journal papers are read show that many readers first look at the illustrations in the Results section before reading the text.
► Ensure a logical flow. If interrupted by material that is too detailed or is not directly relevant, your readers are going to become disoriented and lose the thread.
► Present data in only one way. Do not repeat in the text data that are also presented in a table or graph.
► Limit the amount of detail: You do not need to include every item of data you obtained despite the hard work needed to obtain it. It should not be a blow-by-blow diary of work done. In any piece of research, there inevitably will be results that are not worth presenting.
► Avoid presenting repetitive data. Give representative data, and state that they are representative.

- It is important to include anomalous results that do not support your hypothesis.
- If subheadings are used, they should—if possible—match those used in the Materials and Methods.
- Do not discuss the results. Leave this for the Discussion. Editors will demand a complete separation of the material in the Results and the Discussion sections. If the journal allows it, consider using the useful section "Results and Discussion."
- The Results section is the next easiest section to write, after the Methods section. It is therefore time-efficient to write the Results as soon as you have finished the Methods.

Common faults
- Inadequate textual description. The trends are not pointed out and readers are left to deduce the results from the illustrations.
- Too much detail. Readers do not need every item of data collected.
- Illustrations that are not self-explanatory, due to inadequate subtitling and captioning.
- Repetition in the text of data already shown in the figures and tables.
- Using too many words when citing figures and tables.

Incorrect
It is clearly shown in Table 2 that . . .

Correct
Table 2 shows that . . .
. . . (Table 2).

Tense of the verb
See the Quick Reference Guide: Parts of Speech and Verb Forms (Part 7) for guidelines for using tense in technical documents and definitions and examples of the various tenses of the verb.

Use the past tense. You are describing the results you obtained.

Example
More rupture occurred within the embedded part than in the free zone.

12.3.8   Discussion

Purpose
- To give the answer to the research problem that was stated in the Introduction.
- To explain how the results support the answer.
- To show the relationships among your observations and to place them into the context of other people's work.
- To draw conclusions.

Guidelines for writing it

▶ Describe the significance of the work: principles, relationships, and generalizations.

▶ State your conclusions as clearly as possible.

▶ Discuss the material; do not just restate it.

▶ Point out any exceptions, or any lack of correlation, and define unsettled points.

▶ Show how your results and interpretations agree or contrast with previously published work.

▶ Do not avoid discussing anomalous data; it will be obvious to an expert reader. Be open and honest about inconsistencies or gaps in the data.

▶ Summarize your evidence for each conclusion. Never assume anything.

▶ Keep all your speculation within reasonable bounds.

▶ Do not be afraid to defend your conclusion. But in doing this, treat other studies with respect.

▶ State any limitations of your methods or study design.

▶ State any important implications.

▶ Overall structure for the Discussion: The beginning (state the aim again and briefly summarize the results); the middle (relate your work to that of others, acknowledge anomalies or limitations of your work, form supported conclusions); the end (present the main conclusion and its implications in the last paragraph).

Choice of words

▶ "Prove" is too strong a word. Reviewers prefer the conclusions to be stated less equivocally. In descending order of strength:

These results show/demonstrate	*Very positive.*
These results indicate	*Slightly less strong.*
These results suggest/imply	*Useful if you want to introduce a slight element of doubt or as a politeness if your results contradict a body of evidence.*
These results support	*Useful if you need to demonstrate agreement with a hypothesis or someone else's work.*

"Appear" can be used to avoid sounding too dogmatic:

Thus, XYZ appears to be essential for . . . *sounds positive but much less dogmatic than* Thus, XYZ is essential for . . .

▶ It is acceptable to use hedging words; science is rarely cut-and-dried.

may be; might be; could be; probably; possibly

▶ But don't go to extremes of hedging

Acceptable
These results suggest that A is the cause of B.
These results suggest that A may be the cause of B.

Too cautious
These results suggest the possibility that A may be the cause of B.

Tense of the verb

See the Quick Reference Guide: Parts of Speech and Verb Forms (Part 7) for guidelines for using tense in technical documents and definitions and examples of the various tenses of the verb.

As in the Introduction (see Section 12.3.5), use a mixture of past and present tenses. Use the past tense for results (yours and those of others), but use the present tense for established fact and existing situations.

Common mistakes

- The main point is not clear. A poorly planned Discussion runs the risk of obscuring the main conclusions from the work.
- Excessive length, lack of structure, poor planning, and wordiness. A Discussion should not be an unstructured brain-dump. The points that you want to make should be clearly and briefly expressed.
- The discussion is too short. An assessor may complain that the Discussion is too short and limited. This probably means either that you have not thought out all the implications of your work or that you are not familiar enough with the literature to be able to place your work in context.
- The significance of the material is not discussed adequately. You should not just recapitulate the material; you should place it in the context of other work and draw conclusions from it.
- Conclusions are insufficiently backed up. Unspecified assumptions are made, leading to unjustified conclusions. Each conclusion that is drawn should have a sound basis.
- Some results are ignored.
- The true meaning of data can sometimes be obscured by the interpretation in the Discussion.

12.3.9 Conclusions

A Conclusions section is required by some journals. This is a listing of your conclusions, opinions, and suggestions.

Example

Conclusions

Through this research, the following conclusions can be reached:

No. 1: a strong conclusion.

1. The organoillite modified by cationic surfactant with lower CMC and longer hydrophobic tail group . . . is an effective sorbent for anionic contaminants such as chromate.

No. 2: a suggested application, following on from Conclusion No. 1.

2. Taking the combination of superior retention of anionic contaminants and its still relative low-hydraulic conductivity, illite modified to a surfactant loading beyond its cation-exchange capacity would greatly assist in preventing migration of contaminant anions in applications such as landfill liners.

Adapted from Li, Z., Alessi, D., Zhang, P., and Bowman, R. S. (2002). "Organoillite as a low permeability sorbent to retard migration of anionic contaminants." *Journal of Environmental Engineering*, **128**, (7), 583.

12.3.10 Designing figures and tables for a journal paper

The Instructions to Authors will give details about this. For the conventions governing using illustrations, see Sections 15.4, 15.5, and 15.6.

Checklist for journal and conference papers

Title
☐ Does it identify the information that a reader would need to gain immediate access to the main point of your document?
☐ Does it contain the necessary key information?
☐ Is it too detailed?
☐ Does it make sense? Has the sense been lost during the quest for the minimum number of words?

Abstract
☐ Is your conclusion clearly stated?
☐ Is the story presented logically?
☐ Does it have a beginning (the context), a middle (methods and results), and an end (the conclusions or outcome)?
☐ Does it contain the sort of information that a reader doing a database search would like to find?
☐ Is the description of the methods brief (unless the paper is presenting a new method)?
☐ Do the results make up most of it?
☐ Have nonstandard abbreviations been avoided?
☐ Have citations been avoided?
☐ Have you kept to the word limit? If not, databases may truncate it.

Introduction
☐ Does it adequately review other people's work?
☐ Does it identify the correlations, contradictions, ambiguities, and gaps in the knowledge in this area of research?
☐ Does it, if appropriate, give a historical account of the area's development?
☐ Does it place the study into the context of other people's work?
☐ Does it *clearly* state the purpose of the work in the final paragraphs?
☐ Does it use the past tense for people's findings, and the present tense for established knowledge and existing conditions?

Materials and Methods

☐ Does it give enough information to allow another competent worker in your field to repeat your work?

☐ Does it give the necessary detail about the equipment used, e.g., the model number of an instrument?

☐ Does it avoid detailed description of standard instrumentation and techniques?

☐ Does it give the necessary details of modifications to standard instrumentation and techniques?

☐ Does it give the necessary details of new techniques?

☐ Does it give appropriate details of statistical analysis?

Results

☐ Are the illustrations well chosen?

☐ Are the illustrations well presented and self-explanatory?

☐ Is there an explanatory text pointing out the key results and trends?

☐ Have you avoided giving a blow-by-blow account of the data?

☐ If you have a lot of repetitive data, have you given only representative data in the Results?

☐ Have you avoided discussing the results?

☐ Have you included the results that do not support your hypothesis?

☐ Have you avoided citing references?

☐ Have you used the past tense throughout?

Discussion

☐ Is the Discussion well structured and logical?

☐ Is the main point at the beginning?

☐ Are the other points given in descending order of importance?

☐ Are the results interpreted rather than restated?

☐ Does it show how the results and interpretations agree or contrast with previously published work?

☐ Is each conclusion well grounded? Is there evidence for each one?

☐ Have you avoided far-fetched hypotheses?

☐ Is it free of vague statements?

☐ Is it accurate, fair, and objective regarding other studies' findings?

☐ Is it frank in acknowledging anomalies in your work?

☐ Is it able to explain most anomalies?

☐ Have you used the past tense for results and the present tense for established fact and existing situations?

Conclusions

☐ Is each conclusion fully supported by the material in the rest of the paper?

13

A Conference or Display Poster

The large numbers of participants at conferences result in a lack of time for all of them to give oral presentations of their work. Posters have therefore become a common way of presenting technical information. The standard poster size of about 1 × 1.5 yards does not allow for much information to be presented if one is to avoid the common error of trying to fit in too much. Thus, a poster must by definition be an abstract of the most important points of your work; yet it needs to be both technologically valid and aesthetically pleasing. Because of these constraining characteristics, many poor examples may be seen at conferences. This chapter is designed to help you produce a poster that avoids common mistakes and is intellectually convincing, clear, concise, and visually attractive.

This chapter is directed specifically at producing a conference poster. The general principles also apply to producing a display poster.

Structure of the chapter

13.1 Attending a conference and presenting a poster: the basics

There are usually three items to prepare for a poster presentation at a conference: a conference abstract, a paper, and a poster. They are all different in their requirements.

You will first be asked to submit an abstract of your work. The required length will be from 200 to 300 words and two to three pages. See Chapter 6, particularly Section 6.10.

The abstracts will be used by the conference organizers to decide who will be invited to present their work at the conference.

If you are accepted, you will then be invited to submit your manuscript in the form of a standard journal paper (see Chapter 12). Specific instructions about how to prepare it and a maximum page length will be given.

The papers from all the participants are then collected and published as the conference proceedings. In most cases, the proceedings are published some time after the conference, either as books or CDs.

If you have been asked to present a poster, you will be given the required dimensions and told the time and place in which to display it. You then stand alongside it at the specified time so that people can discuss your work with you.

13.2 Purpose of a poster

Your poster should present the main points of your work as an enlarged graphic display. It should give enough information to inform, but it also should be simple, clear, and creative and should not look too dense, ill-conceived, or sloppy. An effective poster is a blend of selected information and aesthetic design.

13.3 What readers like in a poster

A survey of frequent conference participants showed that they singled out the following eight effective features of posters. This gives some guidelines from which to work.

- A small amount of text, indicating that it will not take long to read.
- Legibility at a distance of 2 yards, that is, a large enough type size with no interference from background color or design.
- An obviously logical layout so that the viewer is led through the material.
- A clear main point.
- An airy layout: not too dense or crammed.
- An interesting title in large type size.
- An interesting use of color.
- Aesthetically pleasing, self-explanatory figures with informative titles and captions.

13.4   Steps in planning a poster

Important: Do not just attach your conference paper and a few illustrations to the board. Some presenters adopt this quickest, least creative method and use the poster as a mounting medium for their paper. The result is poor: too much small text with a dense and squashed appearance, making it overwhelming and discouraging.

An effective poster requires planning of two crucial aspects: the material to include and its layout.

▶ Step 1
Define the main point of your work, the one that you would like people to remember after viewing the poster. This is the so-called "take-home message." In many posters, the main point is not clear.

▶ Step 2
Plan the title. This will need to be decided months ahead of the conference, at the time you submit the initial abstract. Consider how the title will look on a poster. If it is too long, it will take up too much space.

▶ Step 3
To expand the take-home message, *pick no more than a few points* (some authorities say as few as three) and focus on them. You probably will want to put more information into a poster than is realistic. You need to be very selective.

▶ Step 4
Work out how to make the illustrations tell the story. Most viewers look at the illustrations first and only skim the text. The illustrations need to tell the story and show the flow of information. They also need to be simple, with proportionately chunkier lines and larger labeling than would be used in written form.

▶ Step 5
Make sure you know the *size and shape of your poster.* Do not take the dimensions for granted; they will vary from conference to conference. Consider the following:

- The size will determine how much to limit the information; the shape will determine the layout.
- The long axis can be horizontal or vertical.
- The dimensions could be anything from 1 × 1.5 yards upward.
- The length : breadth ratio can vary so that some formats have a longer axis than others.

▶ Step 6

Work out how much *detailed information* is needed. Take into account that the poster viewing session can vary from an hour to the total duration of the conference. You may not be standing by your poster for the entire session. This means that the essential points need to be understood without your explaining them personally.

However, this does not necessitate a dense poster. Too many presenters believe that a poster packed with information gives the impression of productive work. It doesn't. It is likely to obscure the central ideas. Very few people will bother to read every word of a densely packed poster.

Most viewers of a poster want something that looks clear and easy to absorb; however, there will always be a very few people who will argue for including lots of fundamental material. Arguments for limiting the information are:

- You can give out *a handout sheet* with more detail. The people who are really interested in your work will approach you and would appreciate more detailed supplementary information. It is better to have a detailed handout than to try to convey the same degree of detail on the poster alone.
- Studies have shown that poster viewers absorb the information much more readily if *a few points* are clearly stated and well presented rather than given as a mass of detailed information.
- At a conference, the level of lighting may be poor, and the background noise level high. The venue for a poster display may often be a foyer or a convention hall, where conditions may not be ideal for concentrating on detailed material.

▶ Step 7

Plan for the text and illustrations to be easily read from at least 2 yards away, and the title from 5 yards. This means *large type* size, probably larger than feels comfortable. However, you need to tread a fine line between giving visibility with large type size and running the risk of making your work look superficial because of a lack of information. Twenty-two-point type is a good compromise for the main text.

▶ Step 8

Use *story-boarding techniques* to plan the layout. Make sketches on large sheets of paper, or work in the actual size by using a table or the wall and adhesive tape.

▶ **Step 9**

Decide on the ratios of text, graphics, and free space. A rule of thumb that is sometimes used is 20% text, 40% graphics, and 40% free space.

13.5   Design of the layout

It is difficult to give absolute advice about poster design. You need to aim for a well-finished product, with a good balance between text and illustrations. Some suggestions are:

▶ Put the title at the top of the poster, not at the bottom. It will be in large type size and can therefore be read above eye level or from a distance.

▶ Order the progression of information in a manner that is logical and obvious to the viewer.

▶ Plan for the poster to be read from top to bottom and from left to right. Do not make the viewers' eyes jump around. In particular, avoid placing the text first and then filling in the gaps with the illustrations.

▶ Group the information so that related pieces of information are adjacent and the flow to the next item of information is obvious. Arrows can also be useful, but keep the arrangement simple; multidirectional flows are confusing.

In the conventional formal layout, the flow of information is obviously from left to right and from top to bottom (Figure 13-1). The various sections are separated by grouping the text and figures within each one and by leaving space between the groups. The grouping can also be emphasized by the use of color.

If you are aiming for a more eye-catching presentation, take particular care that the direction of information flow is obvious to the viewer.

▶ *If the poster is wide, divide the space into two or more vertical sections.* The viewer will be able to stand in one place and then move to the right to read the next section. It is also convenient when a number of people are reading it simultaneously.

▶ *Make a rough layout of the individual elements.* To visualize the size and position of the individual elements, sketch a rough plan. It should show the approximate positions and sizes of the figures and the individual items of text.

Here are some guidelines for preparing a rough layout:

▶ Decide on your take-home message, your few main points, and how you will build the story around the illustrations.

▶ Sketch an outline that has the scaled-down dimensions of the shape of the poster.

▶ Pencil in different sizes and positions of the elements and ways of fitting this information together.

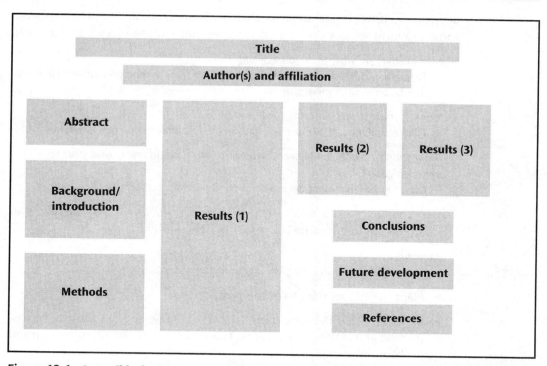

Figure 13-1 *A possible formal layout for a poster, illustrating the flow of information from top to bottom and left to right.*

- Group the information spatially so that the flow of information is clear. The connections between the text and the relevant illustrations also need to be clear. Do not let the information flow disappear in the trial repositionings of the various elements. Later, in the final version, you can enhance the groupings by using color.
- Do not allow the final layout to look as though you have been hunting for space and shoehorning information into gaps.
- Make sure that the illustrations are placed logically in relation to the text of the poster, in particular to the place in the text where each one is cited.
- Do not try to increase the number of illustrations or the amount of text by scaling things down or squeezing them together.
- Be single-minded—don't lose sight of your few main points.
- Your final plan should give a general idea of the amount of text and number of illustrations as well as the approximate positioning and size of text and illustrations.

13.6   Poster title

A poster title needs to contain the key information and to draw the attention of the poster viewers. For this reason, it can be shorter and possibly more querying or controversial than a title for a journal paper. Questions as titles can be provocative, but they also can imply that your results are in question.

Informative, suitable for a journal paper
Strength and modulus degradation of carbon fiber-reinforced polymer laminates in relation to fiber misalignment

Shorter, more direct, suitable for a poster
Fiber misalignment can degrade carbon fiber-reinforced polymer laminates

A question: attracts the viewers' attention, but it could imply that your results are ambiguous
Can fiber misalignment degrade carbon fiber-reinforced polymer laminates?

▶ *If you have used a novel method, show it in the title.* Many people who attend conferences are interested in new methodology.

Example
A new method for assessing ventilation in large spaces

▶ Place the title at the top of the poster.

▶ Make sure that the type size of the title is large enough for it to be easily read at a distance of 5 yards. Titles that are too small are a common problem in posters.

13.7   Possible sections for a poster

Choose your sections to show the flow of information. Your choice of headings is far less constrained than when writing a paper. There also is no need to use the rigid, classic Abstract, Introduction, Methods, Results, and Discussion (AIMRAD) headings. There probably are more logical and attractive ways to present your methods and results.

However, if you do want to use some of the AIMRAD headings, the following list gives some information about how to use them in a poster.

Title
Since its functions are both to inform and to draw the viewers' attention, a poster title should be interesting and does not need to be as formal as that of a journal paper.

Author(s)
In a poster, viewers will often look at posters by specific authors.

Place of work (affiliation)
For the same reason, this is as important as the authors.

Very short Abstract
Some people will maintain that since the text of a poster is short, there is no need for an Abstract. This is not so. An initial overview will always help the reader to better assess your work, whether it is a paper or a poster.

The abstract should be very short (100 words or fewer). Make sure it contains the bare outline of the methods you used and your results. Place it in the logical position where people will expect it: at the top, immediately under the title/authors/affiliation, probably to the left.

Short Introduction *or* Background

This should give the background to your work and include the main references to that of other people.

Methods

- If you have used a standard method, do not describe it in great detail.
- If the main point of your work is to present a novel method, give enough details of it, and also be prepared to discuss it in detail with the viewers of the poster. Also make sure that your title indicates a novel method *(A New Method for . . .)*.

Results

This is best presented as self-explanatory figures with a small amount of text. Avoid tables unless they are absolutely necessary and very simple.

Discussion

You may want a short Discussion section. It may not be necessary.

A brief Conclusions section

It could be in the form of a short list, concisely worded, starting with your major conclusion. Position it at the logical end of the information flow, probably at the bottom right or center.

Future Development

It may be particularly relevant at a conference to show how you are planning to extend this work. However, some people are reluctant to announce the next phase of their research.

References

A References section is used if you need to cite other people's work. It should be a very short section. When space is critical, this section is often printed in smaller type size.

13.8   Figures and tables

Viewers look first at illustrations

Very often, the illustrations are the only part of a poster that people really study. Illustrations should be well presented, clear, and readily understood as far as possible without reference to the text. They should have explanatory titles, clear captions and associated information, and keys to symbols.

Make your illustrations look outstanding

There may be colleagues in your organization with experience using software to produce superb illustrations. Don't be afraid to ask them for advice.

Guidelines for illustrations

▶ Plan the poster around the figures. Use them to tell the story. Well-captioned figures need very little linking text to be able to convey the story line.

▶ Make each figure self-explanatory. A figure with explanatory text in the caption is more convenient for the viewer than one that requires him or her to move from figure to text and back again to be able to understand the argument.

▶ If an illustration has a number, refer to it at the appropriate place in the text. If the reference in the text to the illustration is made prominent (e.g., by being in boldface, upper case, or italics), it is easier for the viewer to cross-refer from text to illustration or vice versa.

▶ Position figures as near as possible to where they are mentioned in the text. Make sure that during the arrangement of all the elements of a poster, the figures do not end up being used to fill gaps; this produces an illogical layout.

▶ Schematic diagrams of equipment and procedures are particularly effective in describing methods.

▶ Graphs must be large, with chunky lines and labeling in large type size. Do not use enlarged copies of your paper's graphs. They will appear too thin and spidery. Limit the amount of information and the number of lines. Make the points, axes, and lines clearly visible from 2 yards away.

Tables

Avoid tables if possible. They usually are not effective on posters unless they are extremely simple. If possible, present the information in other ways. If you decide to use tables, make sure they contain far less information than those in written documents.

Photographs

Make sure that photographs are not enlarged beyond their capabilities. Digital images will become pixelated; other types will be fuzzy and will lack definition.

13.9   Structure of the text

Viewers often scan the text quite rapidly, *so do not use lengthy paragraphs as in your paper.* Use short, simple, and separated statements. Text can be understood more quickly when it has been separated. In the example below, white space and a variable left margin have been created by indenting and listing within the text.

Style as in written text

Sandwich composites are used on aircraft because of excellent stiffness-to-weight ratios. However, they have low damage tolerance and are frequently impacted in normal operation because of their locations in the aircraft. To date, virtually no information has been available on the effects of impact damage.

Short, simple, separated text suitable for a poster

BACKGROUND

Sandwich composites on aircraft:

- have excellent stiffness-to-weight ratios
- have low damage tolerance
- are frequently impacted because of their location

To date, virtually no information has been available on the effects of impact damage.

Every section of text should have a heading to make the flow of information clear. Differentiate the headings from the text by using capital or small capital letters, a bold font, or a different font.

Use plenty of white space in the text. Use indenting of blocks of text and listing within the text to create a variable left margin. Avoid solid blocks of text; they discourage viewers.

13.10 Style of font

▶ **Use a simple font.**

Elaborate fonts are difficult to read and can look unprofessional.

▶ **Don't use too many fonts.**

The usual guideline is no more than two, but use them discreetly. For instance, use one font for the text and one for the headings.

▶ **Use a sans serif font.**

Serif fonts (e.g., Times Roman, Palatino) are commonly used in documents. They usually are regarded as too elaborate to be used in a poster. Moreover, many people consider them old-fashioned.

A sans serif font (e.g., Arial, Helvetica, Century Gothic, Univers) is simple, clean, and easy to read. Some fonts are almost identical. For any one font, letter shape and line spacing may differ so that some fonts appear more dense than others.

Examples

Arial, Helvetica

PROPERTIES
Sandwich composites have excellent stiffness-to-weight ratios.

Helvetica condensed

PROPERTIES
Sandwich composites have excellent stiffness-to-weight ratios.

Century Gothic

PROPERTIES
Sandwich composites have excellent stiffness-to-weight ratios.

Univers

PROPERTIES
Sandwich composites have excellent stiffness-to-weight ratios.

Times New Roman   **PROPERTIES**
Sandwich composites have excellent stiffness-to-weight ratios.

A serif font such as Times New Roman is usually regarded as not suitable because it is considered too elaborate and old-fashioned for posters.

▶ **Bold, uppercase, and italics.**
Avoid blocks of text in UPPER CASE or *italics*; they are difficult to read and discouraging. Use them only in very small amounts for emphasis.

Use boldfacing only for special emphasis, such as the title and headings. Solid text in boldface looks harsh and may be difficult to read.

13.11   Using color and background

Use color inventively. With an intelligent use of color you can unify poster parts by colored backgrounds, make the flow of information easier to follow with color-coded keys and arrows, and make your poster eye-catching. We are visual animals; color attracts our attention.

However, it is absolutely necessary to bear in mind that, when using graphics software, there may be considerable differences between the monitor image and the final printed product. This can depend on both monitor and software characteristics. Many people have been disappointed in the finished product because of this. If possible, it is worth making a test print.

Drawing software or a presentation software package such as Microsoft Power-Point® can produce superb poster material. You can readily produce either individual slides to be attached to a background or a one-piece poster if a large-format printer is used. This can look very professional and is easy to transport. However, beware of some of the presentation software standard backgrounds; they can be fussy and ill-colored. Choose carefully and customize the color or design your own. Moreover, the standard backgrounds are easily recognizable; viewers may get the impression of a hasty job.

13.12   Printing the poster

Allow enough time for things to go wrong with the printing process, particularly if you use a commercial printing organization. Complications can be caused by an unexpected need to convert file types and other factors that can cause myriad forms of chaos if production has been left to the last minute.

13.13   Final production

Consider the following guidelines if your poster is made up of individual components.

► **Trimming**

If the poster is carelessly trimmed, the result will look amateurish. Use a paper cutter or a sharp hobby knife and a metal ruler, and make sure that you are trimming to right angles.

► **Mounting poster components**

Poster board and mat board come in different colors. Poster board is relatively easy to cut; mat board is not. Large sheets of colored paper are also available, but the result can look amateurish because of its lack of stiffness. However, it is easier to transport.

Avoid using a different-colored piece of paper as the background for each individual component of your poster; it may look gauche.

► **Attaching the components to the mounting medium**

To stick the sheets of paper to the mounting board, use a glue stick for very small items. This is difficult to apply evenly to large items, and the result may look lumpy. Sheets of paper can be attached with a spray adhesive; this coats the paper evenly. Avoid rubber cement and contact adhesive in tubes.

► **Final finishing of the individual components**

Laminating of the individual components or of the complete poster looks professional and provides good protection.

► **Attaching the poster to the conference board**

Tacks or drawing pins may look gauche. It may be worth considering using hook-and-loop fastening such as Velcro™ or similar systems to attach individual components of the poster to the mounting board. It can be used either in strips or in sheets large enough to be tacked to the edges of the mounting board. Strips of the corresponding tape are then stuck to the back of the poster components. This gives a rapid method of accurately lining up the components and quickly rearranging them.

13.14   Common mistakes

- Trying to cover too many points, and cramming in far too much information, with too much detail.
- No technological validity because of gaudiness of presentation and/or lack of detail.
- Text size too small: unable to be read from 2 yards.
- Title size too small: unable to be read from 5 yards.

- Color and background that interfere with the readability of the material.
- Unclear flow of information.
- Unclear main points.
- Lack of free space; layout is not airy enough.
- Illustrations too finely drawn and therefore difficult to see.
- Illustrations placed illogically in relation to the text.
- Tables that contain far too much information.
- Photographs enlarged beyond their capabilities.
- An overall lack of visual appeal.

Checklist for a conference poster

- ☐ Have you avoided just attaching your conference paper to a board?
- ☐ Have you planned the poster around the illustrations?
- ☐ Have you avoided trying to present too much information?
- ☐ Does it look as though the information is easily extractable by a viewer?
- ☐ Is the flow of the story self-evident to a viewer?
- ☐ Is there plenty of free space?
- ☐ Is there only a relatively small amount of text?
- ☐ Is the type of a size that is easily readable from 2 yards away? Can the title be read from 5 yards away?
- ☐ Does the color or background interfere with the legibility of the text or illustrations?
- ☐ Is the text in a clearly legible sans serif (Arial, Univers, Avant Garde) or simple serif (Times Roman, Palatino) font?
- ☐ If it is wide, is the poster divided into three or four sections?

Illustrations

- ☐ Are they self-contained, with self-explanatory, informative titles and captions?
- ☐ Can they be understood in overall terms without refering to the text?
- ☐ Are the lines chunky, with clear labeling?
- ☐ Have you avoided squeezing the illustrations into spaces left between the text items?
- ☐ Have you avoided tables if possible? If a table is needed, is it simple?

- ☐ Is color used inventively and intelligently?
- ☐ Have you planned the most professional method of attaching your poster to the viewing board?
- ☐ Is the poster transportable?

Referencing; editorial conventions; and revising, proofreading, and reviewing

Part 4 gives guidelines for some of the meticulous and essential processes that contribute to the professionalism of a document.

It begins by describing the arcane conventions that govern the citation of sources. It also explains the stylistic conventions that are most commonly used in engineering documents. It finishes by offering guidelines for the pedantic and necessary processes of revising the first draft of a document and subsequently proofreading it, together with the reviewing process that ensures the Terms of Reference are correctly addressed.

14

Referencing Your Sources

Formal documents that cite source material need to use either a bibliography or a full referencing system. A Bibliography section—made up of a list of the sources used in the preparation of the document, together with others of interest—is the form that is usually used in the types of formal reports produced by engineering organizations. Journal and conference papers, however, demand the use of a full referencing system. In this, the sources are cited throughout the document, and the full details of each are listed in a section called *List of References* or *References*. All of these systems are governed by traditional conventions. The aim of this chapter is to give full details while assuming no previous knowledge.

Structure of the chapter

14.1 **Purpose of referencing**
14.2 **Referencing a document: the basics**
14.3 **When references should be used**
14.4 **The two main systems of referencing**
14.5 **Personal communications**
14.6 **Sample text and corresponding References section for the two main systems**
14.7 **Using direct quotations with quotation marks**
14.8 **Compiling a Bibliography**
14.9 **Common faults**
Checklist for References

14.1   Purpose of referencing

- To acknowledge other people's work or ideas in relation to your own.
- To enable readers to find the source material.
- To avoid plagiarism, or literary theft. Failure to acknowledge sources is plagiarism and is regarded as a form of stealing. People who do not fully acknowledge their sources or who copy text word for word are implicitly claiming that the work is their own.

14.2   Referencing a document: the basics

A Bibliography is often used in engineering documentation. This is a list of material you have consulted but not cited in the text, together with any other material that you think may be of use to the reader.

Learned engineering and scientific journals or conference papers require a system of referencing based on text citations and a listing of the sources. This system has the following features:

- The sources that you used in preparing your document (papers, reports, books, articles, web sites, etc.) are cited at the appropriate places in the text.
- All the sources are listed together at the end of your document in a References section. This section gives full bibliographic details of each source. It is the final section of the document and is placed immediately before the Appendixes.
- The two basic systems of referencing are the author-date (APA/Harvard) system and the numbering system.

14.3   When references should be used

You need to use references:

- ▶ *When you use factual material taken from other sources.* This is the most common form of citation in an engineering document. The sources may include:
  - *Material on paper,* such as papers in professional journals and conferences; professional reports; books or book chapters; theses; magazine articles; newspaper articles; an organization's publicity material; engineering standards and specifications; government documents, such as Acts of Congress and reports of committees; and other similar sources.

- *Electronic sources* such as web pages, online journal papers, online conference proceedings, CD-ROMs, and electronic databases.
- *Visual and audio material* such as video- and audiocassettes, CDs, and DVDs.

► *When you need to quote word for word from another work* (see Section 14.7). This need for verbatim quoting is not common in technical writing.

14.4   The two main systems of referencing

There are two broad systems commonly used in technical documentation for cross-referencing citations in the text with the full reference in the References section. Here is an overview of the two systems.

Format of the citation in the text of the document

The author/date (APA or Harvard) system
- Surname of the author and the date of publication placed in parentheses. For example: . . . (Brown, 2000).
- Page numbers of a book can be included if needed. For example: . . . (Smith, 2002, pp. 103–121).

The numbering system
- Each citation in the text is given a unique number, either in square brackets, e.g., [5], or superscripted, e.g., [5]. Each is numbered in the order in which it appears in the text.
- If you need to cite a reference more than once in the text, the number of its first appearance (its unique number) is used each time you cite it.

Format of the List of References

The author/date (APA or Harvard) system
- Sources are listed in alphabetical order of the surnames of the authors.

The numbering system
- The sources are not listed alphabetically. It is a list numbered from 1 to *n*, the number of each listing corresponding to the unique number that each source was assigned in the text.

14.4.1   The advantages and disadvantages of the two referencing systems

Scholarly journals and conferences will always specifically state the system required to be used. However, if you are free to choose, here is a guide to the advantages and disadvantages of each system.

Advantages

The author/date (APA or Harvard) system

- Allows the source to be recognized by author and date in context within the text of the report. This is seen as a considerable advantage by people familiar with the literature.
- Provides an alphabetical list at the end of the document.
- Allows easy insertion of an extra reference into the text.

The numbering system

- The text of the document is not interrupted by lengthy author/date citations.
- Only a number needs to be repeated, which prevents repetition in the text of the same wordy citations.

Disadvantages

The author/date (APA or Harvard) system

- Can create disruption in the text when there are many author/date citations in one place.

The numbering system

- While reading the text, readers familiar with the literature cannot recognize the work that you are citing. They have to turn to the List of References to match a numerical reference to its source.
- It may be difficult to add another citation and renumber all successive ones. This can be overcome by using the word processor's endnoting function or a referencing software package.
- The numbers give no information about the work, and it is easy to forget to use the earlier number when you need to refer to it again later in your report. Again, the word processor's endnoting function or a referencing software package will overcome this.

Important: Either one system or the other must be used in a document. Care must be taken not to use a mixture.

14.4.2   Citing the sources in the text of the document

The author/date (APA or Harvard) system: text citations

Overview: The sources cited in the text are in the form of (author, year of publication). Here are examples of text citations using the author/date system for various types of document.

- Author's surname and date placed after the information.

   A rigorous mathematical formulation of this behavior has been based on plasticity (Chen, 1994).

- Author surname(s) given in the text.

Follansbee and Frantz (1983) showed that high frequency dispersion overriding the main wave is inevitable.

- Two authors.

  Average root-zone salinity has been stated to be an approximate indicator of crop growth in irrigated areas (van Genuchten and Hoffmann, 1984).

- More than two authors. Cite the surname of the first author and add "et al." (italicized in some house styles).

  The average percent air voids has been related to fatigue cracking (Tayebali et al., 1994).

  or

  Tayebali et al. (1994) showed that the average percent air voids is related to fatigue cracking.

- Several sources cited within one set of parentheses. Separate them by semicolons and, depending on house style, cite them in order of either publication date or alphabetically by author

  A great deal of work has been focused on polymer-modified asphalt (Kraus, 1982; Lee and Denuriel, 1987; Little et al., 1987; King et al., 1992; Lewandowski, 1994; Bandyopadha et al., 1997; Gahvari, 1997; Isaacson and Lu, 1999).

  or

  A great deal of work has been focused on polymer-modified asphalt (Bandyopadha et al., 1997; Isaacson and Lu, 1999; Gahvari, 1997; King et al., 1992; Kraus, 1982; Lee and Denuriel, 1987; Lewandowski, 1994; Little et al., 1987).

- Two or more papers written in different years by the same author.

  Westergaard's well-known equations for stresses in concrete slabs (Westergaard, 1926, 1948) are based on Winkler's assumption.

- Several papers written in one year by the same author. Distinguish between them by adding a lowercase letter to each paper; these letters must be added to the citation dates in the List of References section.

  Previous analysis of this phenomenon (Brown et al., 2000 a, b) has shown that. . . .
  In one specific instance (Brown et al., 2000 b), it was found that. . . .

- Large body of work, but only a few representative examples are cited. Use e.g. within the parentheses.

  It is well established in the literature that air void distribution is a major factor that affects the performance of asphalt mixtures (e.g., Monismith, 1992).

- Large body of information contained in a review paper.

  Filtration of water has been extensively studied (for review, see Brown et al., 2000).

- Original reference unavailable, but cited in another paper. Cite the secondary source and include the primary source. Include full citation details of both references in the List of References section.

  Smith (1928) as cited by Brown (2001) showed that. . . .

- Different authors with the same surname, publishing in the same year.

  It has been shown by Smith, C. W. (1998) . . .

  whereas Smith, J. G., (1998). . . .

- References precisely placed.

  This runoff has also introduced heavy metals (Jackson, 1994), pesticides (Seidel, 1995), pathogens (Brown, 1999), sediments (Fenwick, 1991), and trash (Milner, 2001).

- Publication date of the source known only approximately. Use a small c. ("circa") before the date.

  All the branches of a tree at any degree of height, if put together, are equal to the cross section of its trunk (Leonardo da Vinci, c. 1497).

- Author not stated. Use the first few words of the title, and the date if known. For example, where the citation is:

  — Wylie Stream Intake Feasibility Report (2004). James Consultants Ltd., Contract TKA 97/101. Prepared for Middletown Central Electricity Generation.

    . . . as shown in a previous study (Wylie Stream Intake, 2004).

  — *CORINAIR Working Group on Emission Factors for Calculating 1990 Emissions from Road Traffic,* 1 (1993). Commission of the European Committees (Office for Official Publications, Luxembourg).

    . . . in accordance with a previous study (CORINAIR Working Group, 1993).

  — Twintex TPP fact sheet (undated). Verdex International S.A.

    . . . as specified (Twintex TPP, undated).

Copying or adapting illustrations

Note: For each of these cases, it will be necessary for copyright reasons to obtain permission from the copyright holder of the original source.

- Exact copy of an illustration from someone else's work

  Figure 4 Comparison of theoretical and experimental grading curves (Reproduced from Fukumoto, 1992)

- Illustration redrawn from someone else's work

  Figure 2.2 Influence of compressibility on uncemented, unaged, predominantly quartz sands (Redrawn from Jamiolkowski, 1985)

- Someone else's data or figure adapted and incorporated into a table or figure of your own

  Figure 3.5 Creep tests on dry and wet pumice sand under a sustained loading of 650 kPa (Adapted from Tsopani, 2000)

The numbering system: text citations

- Each source cited in the text is given a unique number, in the order in which each is cited.

- The unique number is usually placed either in square brackets or super-scripted.
- If you need to cite a reference more than once in the text, the number of its first appearance—its unique number—is used each time you cite it.

The wind velocity and behavior of a geographic region is a function of altitude, season, and hour of measurement [1]. Brown [2] has analyzed changes in sulfur dioxide and sulfate concentrations in air during the period 1985–2002. It was also shown [1] that....

The wind velocity and behavior of a geographic region is a function of altitude, season, and hour of measurement.[1] Brown[2] has analyzed changes in sulfur dioxide and sulfate concentrations in air during the period 1985–2002. It was also shown[1] that....

14.4.3   Compiling the List of References

The References section is made up of a list of the papers, books, articles, and other sources that you have cited in the text of your work. It is placed at the end of your document, immediately before the Appendixes. In the citation you need to provide the information that will allow another person to retrieve the material.

- *For the author/date system,* the sources are listed in alphabetical order of the surname of the author, or first author if there is more than one.
- *For the numbering system,* the sources are in the form of a sequentially numbered list, the numbers corresponding to the unique number that each source was assigned in the text.

Here are guidelines for listing the various types of sources in the List of References section. A generalized scheme is shown here. Be aware that there may be minor variations in order and formatting of the individual items; this will depend on the house style of the organization or journal.

Books

Information needed

- Surname and initials of the author(s) or editor(s) (surname first, followed by the initials). If editor, place "ed." after the initials.
- The year of publication.
- Title of the book, often italicized.
- If there is a subtitle, it is separated from the main title by a colon (:). (see first example 1, below).
- Title of series, if applicable.
- Volume number or number of volumes, if applicable.
- Edition, if other than the first.
- Publisher.
- Place of publication (city or town).
- Page numbers of the material quoted (if applicable).

Examples of citations of books in the List of References section

- One or more authors

  Gann, D. M. (2000). *Building innovation: Complex constructs in a changing world.* American Society of Civil Engineers (Thomas Telford, Ltd.).

- One volume of a multivolume work

  Lay, M. G. (1998). *Handbook of road technology*, Vol. 2. Gordon and Breach Science Publishers, New York.

- Second or later edition of the book

  Merritt, F. S., and Ricketts, J. T. (eds.) (2000). *Building design and construction handbook*, 6th Edition. McGraw-Hill, New York.

- A chapter or an article in an edited book
  — The chapter title is enclosed in quotation marks
  — The name of the book is preceded by "In:"
  — Denote the editor(s) by ed. or eds.

  Brown, I. D. (1995). "Accident reporting and analysis." In: *Evaluation of human work.* Wilson, J. R., and Corlett, N. E., eds. Taylor and Francis, Bristol, Penn., pp. 969–992.

Journal papers

Information needed
- Surname and initials of the author(s).
- The year of publication in parentheses ( ).
- Title of the paper in quotation marks.
- The name of the journal in its accepted abbreviated form. *Note:* The standard abbreviations for the journals are available in handbooks; see Part 7.
- The volume number of the journal (with the issue number, if there is one, in parentheses).
- The numbers of the pages on which the paper begins and ends. *Note:* The actual page from which your information is taken is not cited.

Examples of citations of journals in the List of References section

- Single author.

  Kholmyansky, M. M. (2002). "Mechanical resistance of steel fiber reinforced concrete to axial load." *Journal of Materials in Civil Engineering,* 14(4), 311–319.

- Two authors.

  Dutta, A., and Mander, J. B. (2001). "Energy based methodology for ductile design of concrete columns." *Journal of Structural Engineering,* 127(12), 1374–1381.

- Multiple authors.

  Sinha, K. C., Bullock, D., Hendrickson, C. T., Levinson, H. L., Lyles, R. W., Radwan, A. E., and Li, Z. (2002). "Development of transportation engineering research, education, and practice in a changing civil engineering world." *Journal of Transportation Engineering*, 128(4), 301–313.

- Paper in the proceedings of a conference. As for a journal paper but in addition: State the number of the conference, its title theme, where it was held, and the date.

  Tawresey, J. G. (1996). "Professional liability—an approach that works." Proceedings of the 14th Structures Congress, *Building an International Community of Structural Engineers*, Vol. 2, Apr. 15–18, Chicago, Ill., USA, pp. 1288–1295.

- Paper in language other than English, not translated. Put (in *language*) at end of the citation. The title may remain in the original language or be translated into English.

  Müller, R. (1955). "Wasserfassungen in geschiebeführenden Flüssen." *Wasser- und Energiewirtschaft* 9, 11–13 (in German).

  Orlov, D. L., and Gorin, A. E. (1999). "Use of glass fiber concrete in structural engineering." *Meditsinskaya Tekhnika* 3, 3–7 (in Russian).

Other types of sources

Information needed

- If the author is not stated, describe the source as fully as possible in the style of the relevant examples below. The order of the items cited is:
  — The title of the document
  — Date (when possible)
  — The organization/institution that produced the document
  — Any identifying number, such as designation code, or contract number

- For citation in the text of a source with no stated author, when using the author/date system, use an abbreviated form of the title.

Examples of citations of various types of sources in the List of References section

- CD article, video- or audiocassette: Distinguish between CD, videocassette, or audiocassette.

  Engineering Disasters. (1999). A&E Television Networks. Catalogue number AAE 42337. CD.

- Codes.

  "Building Code Requirements for Structural Concrete (ACI 318-99) and Commentary (ACI 318R-99)" (1999). ACI Manual of Concrete Practice, 2000, Part 3. *Use of Concrete in Buildings—Design, Specifications, and Related Topics.* ACI International, Farmington Hills, Mich.

- Consulting report: Include name of consulting firm, contract number, and for whom the report was prepared.

  "Wylie Stream Intake Feasibility Report" (1997). James Consultants Ltd., Contract TKA 99/136. Prepared for Middletown Central Electricity Generation.

- Fact/data sheet: no author, undated.

  "Twintex TPP fact sheet" (undated). Verdex International S.A.

- Legal material.

  See *The Bluebook: A uniform system of citation.* (2000). 17th Edition. Harvard Law Review Association, Cambridge, Mass.

- Map.

  Yellow Water Reservoir (1979). 46° 52' 30", 108° 22' 30". Map No. TMT3390. Scale 1: 24 000. US Geological Survey, Reston, Va.

- Microfiche.

  Serff, N., Seed, H. B., Makdisi, F. I., and Chang, C-Y. (1976). "Earthquake induced deformations of earth dams." Report No. EERC 76-4. Earthquake Engineering Research Centre, Richmond, Calif. Microfiche.

- Newspaper or magazine article: If no author is stated, put title of article in first place. If possible, state the source (e.g., Reuters, IPA, etc.).

  Murray, S. (2002). "Bill overhauling audit regulation passes in Senate." *The Wall Street Journal,* Tuesday, July 16, 2002. p. A3.

- Patent.

  Daudet, L. R., Ralph, G. S., and Ponko, E. L. (2002). "Floor system and floor system construction methods." U.S. Patent number 6,418,694.

- Report by a professional body.

  "Recycling Household Waste—The Way Ahead" (1991). Association of Municipal Engineers, The Institution of Civil Engineers, London.

- Standard practice.

  Standard Practice for Sampling Freshly Mixed Concrete (1999). Designation C 172-99. American Society for the Testing of Materials, Annual Book of ASTM Standards 2001, **04.02**, *Concrete and Aggregates*, 106–108.

- Standard specification.

  ASTM. (1989). "Standard Specification for Plain and Steel-Laminated Elastomeric Bearings for Bridges." *D 4014-89*. American Society for the Testing of Materials, Annual Book of ASTM Standards 2001, **04.03**, *Road and Paving Materials; Vehicle-Pavement Systems*, 401-406.

- Standard test method.

  ASTM. (1997). "Standard Test Method for Performance of Wood and Wood-Based Floor and Roof Sheathing under Concentrated Static and Impact Loads." *E 661-88* (Reapproved 1997). American Society for the Testing of Materials, Annual Book of ASTM Standards, 2001, **04.11**, *Building Construction*, 194–200.

- Technical report.

  Pekcan, G., Mander, J. B., and Chen, S. S. (2000). "Seismic retrofit of end-sway frames of steel deck-truss bridges with a supplemental tendon system: Experimental and analytical investigation." Technical Report No. MCEER-00-004. Multidisciplinary Center for Earthquake Engineering Research, University at Buffalo, N.Y.

- Thesis.

   Loveluck, P. (2004). "Modeling pile behavior under lateral cyclic loading." PhD thesis, Department of Civil Engineering, University of Middletown.

- Undated documents: Put "(undated)" where the date is normally placed.

   "Predicting traffic accidents from roadway elements on urban extensions of state highways" (undated). Bulletin 208, New Zealand Highway Research Board.

Electronic references

Online sources are not regarded as being permanent in the same way as are sources on paper. The conventions therefore require that you state the date of your retrieval of a source to show that it existed at that time.

Information needed
In general terms:
- Author surname(s), initials (if authored). If no author is listed, lead with the title.
- Date (in parentheses), if stated.
- Title.
- Retrieved *month day, year,* from URL.

Examples of citations of various types of electronic references in the References

- Abstract from databases.

   El-Din, A. G., and Smith, D. W. (2002). "A neural network model to predict the wastewater inflow incorporating rainfall events." *Water Research*, 36(5), 1115–1126. Retrieved October 3, 2002, from Compendex database.

- Online journal article.

   Park, S., Kim, Y-B., and Stubbs, N. (2002). "Nondestructive damage detection in large structures via vibration monitoring." *Electronic Journal of Structural Engineering*, 2, 59–75. Retrieved September 15, 2002, from http://www.civag.unimelb.edu.au/ejse/.

- Paper from online conference proceedings.

   Wu, F., and Chang, F.-K (2001). "A built-in active sensing diagnostic system for civil infrastructure." *Proceedings of SPIE—The International Society for Optical Engineering, Smart Systems for Bridges, Structures, and Highways-Smart Systems, Structures and Materials,* March 5–7, 2001, Newport Beach, Calif., 4330, 27–35. Retrieved September 18, 2002, from Compendex database.

- Web page, authored.

   Berg, T. W. (2002). *Fiber Reinforced Concrete.* Retrieved August 30, 2002, from http://www.retailsource.com/information/fiber_rc/fiber_rc.html.

- Web page, no stated author.

  Access Watch Version 2. Retrieved October 9, 2002, from http://accesswatch.com/.

- Online newspaper.

  Grimsley, K. D. (2002). "Workplace changed by telecommuting." *The Washington Post*, August 31, 2002; Page E1. Retrieved September 5, 2002, from http://www.washingtonpost.com/.

- Personal e-mail: Because these are personal communications and cannot be readily retrieved by the general public, most authorities believe that no entry should appear in the References section.

  Instead, *either* acknowledge it in the text in parentheses:

  The testing was discontinued after a week. (H. R. Jones, e-mail to author, April 2, 2004).

  or as a personal communication:

  The testing was discontinued after a week. (H. R. Jones, *pers. comm.*).

14.5   **Personal communications**

This term refers to information given to you personally (e.g., in discussion or by letter, e-mail, or fax). Personal communications usually are cited only in the text as (Initials, Surname, pers. comm.) and are not included in the List of References.

Example
The sample was maintained at 40°F (D. J. Wilson, *pers. comm.*).

However, if you have a number of personal communications, to give them authenticity, it may be appropriate to have a separate section for them, after the References, headed *List of Personal Communications*. They should be listed in alphabetical order of the surnames. State where they work. You may also want to include the means of communication and the date.

Example
List of Personal Communications
1. Adams, B. A., Department of Civil Engineering, University of Technology, Middletown. By letter, 8/7/2004.
2. Wilson, D. J., Department of Environmental Science, University of Middletown. In discussion, 5/7/2003.
3. *[etc.]*

Alternatively, you may prefer not to use the personal communication system and refer to these people in the Acknowledgments section.

14.6 Sample text and corresponding References section for the two main systems

Author/date (APA or Harvard) system

Commentary	Text

Electronic sources: Four with cited authors, one with no cited author ⟶

The recent upsurge of interest in the mechanical efficiency of medieval hurling devices has resulted in their use as student construction projects in engineering (O'Connor, 1994). There is also a wealth of Web-based material: for instance, graphics, and information (Miners, 2000), desktop models (Toms, 2002), and computer simulations of trebuchets (Siano, 2001; *The virtual trebuchet*, 1998).

Repeat of a previously cited reference

Author mentioned in text ⟶

Three references in a series, placed in chronological order, separated by semicolons

More than two authors; uses "et al."

Precise placing of references in the text, one referring to the palintonon, and another to the onager

Used in ancient times to hurl everything from rocks to plague-ridden carcasses of horses (O'Leary, 1994) and—in a modern four-story-high reconstruction—dead pigs, Hillman cars, and pianos (O'Connor, 1994), the trebuchet relied on the potential energy of a raised weight. Its mechanical efficiency has been compared unfavorably by Gordon (1988) with that of the palintonon, the Greek hurling device, which could hurl 40 kg stone spheres over 400 meters (Hacker, 1968; Marsden, 1969; Soedel and Foley, 1979). This device incorporated huge twisted skeins of tendon, a biomaterial that can be extended reversibly to strains of about 4% (Wainwright et al., 1992). The palintonon used the principle of stored elastic strain energy, the fact that when a material is unloaded after it has been deformed, it returns to its undeformed state due to the release of stored energy (Benham et al., 1996). The motion of the palintonon (Hart, 1982) and that of its Roman equivalent, the onager (Hart and Lewis, 1986), has been analyzed by use of the energy principle applied to the finite torsion of elastic cylinders.

References

Book: Note publisher, place of publication (Harlow), and relevant page(s)

Benham, P. P., Crawford, R. J., and Armstrong, C. G. (1996). *Mechanics of Engineering Materials*, 2nd Edition. Longman, Harlow, p. 67.

Book

Gordon, J. E. (1978). *Structures or why things don't fall down*. Penguin, Harmondsworth, pp. 78–89.

Chapter in book; book is Volume 9 of a series

An "In:" reference

Hacker, B. C. (1968). "Greek catapults and catapult technology: Science, technology and war in the ancient world." In: *Technology and Culture*, 9, pp. 34–50.

Paper in journal

Hart, V. G. (1982). "The law of the Greek catapult." *Bull. Inst. Math. Appl.*, 18, 58–68.

Paper in journal	Hart, V. G., and Lewis, M. J. T. (1986). "Mechanics of the onager." *J. Eng. Math.*, 20, 345–365.
Book	Marsden, E. W. (1969). *Greek and Roman artillery*. Clarendon Press, Oxford, pp. 86–98.
Electronic source with cited author	Miners, R. (2002). "The Grey Company Trebuchet Page." Retrieved October 4, 2006, from http:// members. iinet.net.au/~rmine/gctrebs.html.
Article in journal, no volume number	O'Connor, L. (1994). "Building a better trebuchet." *Mechanical Engineering*, January, 66–69.
Editorial in journal	O'Leary, J. (1994). "Reversing the siege mentality." *Mechanical Engineering*, January, 4.
Electronic source with cited author	Siano, D. (2002). "The algorithmic beauty of the trebuchet." Retrieved October 8, 2006, from http://members.home.net/dimona/.
Electronic source, no cited author	"The Virtual Trebuchet" (2002). Retrieved October 8, 2006, from http://www.stud.ifi.uio.no/~oddharry/ blide/vtreb.html.
Electronic source, authored	Toms, R. (2002). "Ron L Toms' Products and Services." Retrieved October 4, 2006, from http://www.rlt.com/.
Article in magazine with volume number in boldface	Soedel, W., and Foley, V. (1979). "Ancient catapults." *Scientific American*, **240**, 150–160.
More than two authors; an "et al." reference in the text	Wainwright, S. A., Biggs, W. D., Currey, J. D., and Gosline, J. M. (1992). *Mechanical design in organisms*. 2nd Edition. Longman, Harlow, p. 83.

Numbering system

Commentary	*Text*
	The recent upsurge of interest in the mechanical efficiency of medieval hurling devices has resulted in their use as subjects for student construction projects in engineering [1]. There is also a wealth of web-based material: for instance, graphics and information [2], applications such as desktop models [3], and computer simulations of a trebuchet [4, 5].
A second reference to Source Number 1; it is not assigned a new number ⟶	Used in ancient times to hurl everything from rocks to plague-ridden carcasses of horses [5] and, in a modern four-story-high reconstruction, dead pigs, Hillman cars, and pianos [1], the trebuchet relied on the potential energy of a raised weight. Its mechanical efficiency has
Author mentioned in text ⟶	been compared unfavorably by Gordon [6] with that of the palintonon, the Greek hurling device, which could
Three references in a series, separated by commas ⟶	hurl 40 kg stone spheres over 400 meters [7, 8, 9]. This device incorporated huge twisted skeins of tendon, a biomaterial that can be extended reversibly to strains of

Precise placing of references in the text; one referring to the palintonon, and another to the onager ➤

about 4% [10]. The palintonon utilized the principle of stored elastic strain energy—the fact that when a material is unloaded after it has been deformed, it returns to its undeformed state due to the release of stored energy [11]. The motion of the palintonon [12] and that of its Roman equivalent, the onager [13], has been analyzed by use of the energy principle applied to the finite torsion of elastic cylinders.

References

1: Article in journal, no volume number

1. O'Connor, L. (1994). "Building a better trebuchet." *Mechanical Engineering,* January, 66–69.

2, 3, 4: Electronic sources, each with a cited author

2. Miners, R. (2002). "The Grey Company Trebuchet Page." Retrieved October 4, 2006, from http://members.iinet.net.au/~rmine/gctrebs.html.

3. Toms, R. (2002). "Ron L Toms' Products and Services." Retrieved October 4, 2006, from http://www.rlt.com/.

4. Siano, D. (2002). "The algorithmic beauty of the trebuchet." Retrieved October 8, 2006, from http://members.home.net/dimona/.

5: Electronic source, no cited author

5. "The Virtual Trebuchet" (2002). Retrieved October 8, 2006, from http://www.stud.ifi.uio.no/~oddharry/blide/vtreb.html.

6: Book: Note publisher, place of publication, and relevant page number(s)

6. Gordon, J. E. (1978) *Structures or why things don't fall down.* Penguin, Harmondsworth, pp. 78–89.

7: Editorial in journal

7. O'Leary, J. (1994). "Reversing the siege mentality." *Mechanical Engineering,* January, 4.

8: Article in magazine with volume number in boldface

8. Soedel, W., and Foley, V. (1979). "Ancient catapults." *Scientific American,* **240**, 150–160.

9: Chapter in book

9. Hacker, B. C. (1968). "Greek catapults and catapult technology: Science, technology and war in the ancient world." In: *Technology and Culture*, 9, pp. 34–50.

10: Book

10. Marsden, E. W. (1969). *Greek and Roman artillery.* Clarendon Press, Oxford, pp. 86–98.

11: Book with four authors

11. Wainwright, S. A., Biggs, W. D., Currey, J. D., and Gosline, J. M. (1992). *Mechanical design in organisms.* 2nd Edition. Longman, Harlow, p. 83.

12: Book with three authors

12. Benham, P. P., Crawford, R. J., and Armstrong, C. G. (1996). *Mechanics of engineering materials.* 2nd Edition. Longman, Harlow, p. 67.

13: Paper in journal

13. Hart, V. G. (1982). "The law of the Greek catapult." *Bull. Inst. Math. Appl.,* 18, 58–68.

14: Paper in journal

14. Hart, V. G., and Lewis, M. J. T. (1986). "Mechanics of the onager." *J. Eng. Math.,* 20, 345–365.

14.7   Using direct quotations with quotation marks

A technical document very rarely uses direct quotes set in quotation marks. If they are needed, they should be very brief indeed—no more than a few words.

Conventions for direct quotations

- *Where the series of words is the same as in the original*

  T. H. Huxley (1825–1895) said that science is "nothing but trained and organized common sense."

  It has been said[4] that science is "nothing but trained and organized common sense."

- *Where very slight changes are needed to a quotation so that it fits your prose*
  For example, a capital letter may need to be changed to lower case, or a noun substituted for a pronoun, or a noun or phrase inserted, so that it makes more sense. These changes are indicated by brackets [ ]. The whole quote is enclosed in quotation marks.

  Steven Jay Gould (1985) has stated that "the history [of human races] is largely a tale of division—an account of barriers and ranks erected to maintain the power and hegemony of those on top."

  It has been stated[2] that "the history [of human races] is . . . etc."

  (The original quote was: *"The history is largely ...."*)

- *Where part of a quote needs to be omitted because it is irrelevant to your document*
  Use three dots (an ellipsis) to show the omission. It is important that the sense of a quotation is not altered by the omission. The whole quote is enclosed in quotation marks.

  Inkster (1991) has said that "the role of technological change . . . may be exaggerated but it may also be underestimated."

  It has been said[6] that "the role of technological change . . . etc."

14.8   Compiling a Bibliography

The conventions used for compiling a Bibliography are as follows:
- Each listed citation is formatted as for the References (see preceding).
- The items are listed in alphabetical order.
- The list is not numbered.

It is common practice to indent each line of a reference after the first. Use the hanging indent function on a word processor (Ctrl T on Microsoft Word®).

The following example of a Bibliography includes sources on paper and electronic sources.

Bibliography

Benham, P. P., and Crawford, R. J. (1987), *Mechanics of engineering materials.* Longman Scientific and Technical, Harlow, pp. 66–68.

Gordon, J. E. (1978). *Structures or why things don't fall down.* Penguin, Harmondsworth, pp. 78–89.

Marsden, E. W. (1969). *Greek and Roman artillery.* Clarendon Press, Oxford, pp. 86–98.

O'Connor, L. (1994). "Building a better trebuchet." *Mechanical Engineering,* January, 66–69.

Siano, D. (2002). *The algorithmic beauty of the trebuchet.* Retrieved October 8, 2004, from http://members.home.net/dimona/.

The Virtual Trebuchet (2002). Retrieved October 8, 2004, from http://www.stud.ifi.uio.no/~oddharry/blide/vtreb.html.

Wainwright, S. A., Biggs, W. D., Currey, J. D., and Gosline, J. M. (1976). *Mechanical design in organisms.* Princeton University Press, Princeton, N.J., pp. 88–93.

14.9   Common faults

- A reference is cited in the text and omitted from the References, and vice versa.
- The date of the text citation does not correspond to that of the citation in the References.

These two faults tend to be regarded as unforgivable by many assessors.

- In the text citations: Sentences are broken up by frequent, long strings of citations.
- In the List of References:
  — The formatting is inconsistent.
  — A nonstandard abbreviation for a journal is used.
  — Insufficient details are given: in particular, omitting the publisher and place of publication of a book.
- Citing references that are unobtainable.
- Volume and page numbers are incorrect.

Checklist for References

Remember, references are needed when you cite factual material from the literature and you quote directly from another work. Decide whether you need a References or Bibliography section. Most publishers prefer a List of References section.

☐ For each one of your text citations, is there a corresponding reference in the List of References? And vice versa?

☐ Does the date of the text citation match the date in the full reference in the References?

☐ Are all the references in the References formatted consistently?

☐ Are all the necessary details given in the List of References?

15

Editorial Conventions

Technical documents have particular conventions of editorial style. This chapter describes an acceptable form for the most common editorial conventions that cause confusion. For further information, refer to a style manual (see Part 7).

Structure of the chapter

15.1   Conventions for writing numbers in the text

- Measured quantities

  Figures   2.4 seconds, $5,000, 50°F, 9 miles, 6 tons, 3 amps

- Counted numbers
  - One to nine, words   … in five areas
  - More than nine, figures   … in 11 areas

- Number at the beginning of a sentence

  Words, regardless of size   Two hundred and forty samples were taken.

- Ordinal numbers

  *Same as for counted
  numbers*   First, second, third ... 11th ...

- A series of numbers above and below 10

  Figures   ... over periods of 3, 6, and 12 hours

- Percentages

  Figures   ... that 8 percent of the samples ...

- Fractions (but better expressed as a percentage)

  Words   ... one-fifth of the surface area

- Dates and times

  Figures   ... on July 8, 2006

  ... at 8.30 a.m. (or 08:30)

For dates, use the format: Month (written out) Day (in figures) Year in figures (with a comma between day and year).

Correct   July 8, 2006

Incorrect   7/8/06 (different countries use different formats
      when using only figures; it may cause confusion)
      8 July 2006
      8th July 2006 (regarded as old-fashioned)

- Reference in the text to figures and tables

  Figures   Figure 3 shows that ...

  ... (Table 2)

15.2   Rules for capitalization

15.2.1   Titles of books and journals

The general convention is to capitalize the initial letters of the "main" words in the titles, i.e., the words other than small words such as articles, prepositions, conjunctions, and so on (*see* The Quick Reference Guide, Parts of Speech and Verb Form in Part 7, for the meanings of these definitions). The correct term for this is *title case*.

Common words that do not have initial capitals unless they are the first word in the sentence:

Articles
    the   a

Prepositions
across	up
by	down
for	of
to	

Conjunctions
and	because
but	for
so	although
since	

Coordinators
if . . . then	neither . . . nor
both . . . and	whether . . . or
either . . . or	

Examples

Building Innovation: Complex Constructs in a Changing World

Journal of Professional Issues in Engineering Education and Practice

15.2.2   Titles of journal articles; figure and table captions

Use capitals only for the first word. The correct term for this is *sentence case*.

Mechanical resistance of steel fiber-reinforced concrete to axial load

Figure 4 Generalized relationship between particulate air quality in major urban areas of the world and per capita income (adapted from Lomborg, 2001)

15.2.3   Referring in the text to figures, tables, chapters, rows or columns of tables, or pages

Figures, tables, chapters, sections

▶ Use initial capitals when referring to specific figures, tables, chapters, or sections.

... is shown in Figure 3.
... as given in Table 2.
... is described in Chapter 6.
... is analyzed in Section 5.

Rows or columns of tables, pages

▶ Do not use initial capitals when referring to rows or columns of tables or to pages.

... as given in row 2 of Table 12.
... as given in column 3 of Table 12.
... see page 38.

15.3   Defining acronyms in the text

Acronyms are in the form of the initial capital letters of a series of words. Examples are polymer modified asphalt (PMA), Californian bearing ratio (CBR), and volatile organic compounds (VOC).

The convention is that the terms are spelled out at their first use in the text, followed by the acronym in parentheses.

> The vapor pressures of many volatile organic compounds (VOC) increase with temperature.

In the remainder of the text, you may then use only the acronym.

However, because of the possible difficulty in finding the first use of the term in the text, it helps the reader if—as well as being defined in the text—abbreviations are included in a section called Glossary of Terms and Abbreviations (see Section 5.3.7).

15.4   Numbering of chapters and sections of documents, pages, and illustrations

15.4.1   Chapters and sections of documents

This section describes the conventions for the decimal point numbering system for numbering chapters or sections of a document and their associated subheadings and subsubheadings.

The main chapters/sections are given Arabic numerals. The subsections are denoted by putting a decimal point after the section number and another Arabic numeral:

1.0   Title of first main chapter/section
    1.1   First subheading
    1.2   Second subheading

2.0   Title of second main chapter/section
    2.1   First subheading
    2.2   Second subheading
        2.2.1   First division in the second subheading
        2.2.2   Second division in the second subheading
        2.2.3   Third division in the second subheading
    2.3   Third subheading

3.0   Title of third main chapter/section

15.4.2   Pages

Smaller documents
The conventions associated with page numbering are:
- Page 1 is the first page of the Introduction.

- All the preliminary pages—i.e., those before the Introduction (Title Page, Executive Summary, Acknowledgments, Table of Contents, List of Illustrations, Glossary of Terms and Abbreviations, etc.)—are assigned lower-case Roman numerals (i, ii, iii, iv, v, etc.).
- The first page that is counted is the Title Page, but it is not labeled as such; it is left blank.
- Each of the other preliminary pages (starting at page ii) is labeled with its number.

Larger documents or documents written by a variety of people or groups

- The pages may be numbered according to the chapter or section (e.g., 4.1, 4.2, . . . 4.12, then 5.1, 5.2 . . .). This has several advantages: it avoids having to renumber the whole document whenever you insert a new page of text; collation is easier when separate chapters or sections are prepared by different people or groups; and the use of preprepared modules is made easier.

Appendixes

- Appendixes can be either numbered or lettered: Appendix 1 or Appendix A.
- The page numbers of the Appendixes are separate from those of the main body of the document and are related to the numbering of the Appendix. For example: Page 1-1, 1-2, 1-3, etc. *or* page A-1, A-2, A-3, etc.

15.4.3 Illustrations

Every illustration (figure or table) in a document must have a number and a title and must be referred to at an appropriate place in the text.

There should be two numbering series: one for all the figures (i.e., everything that is not a table—graphs, maps, line drawings, flow diagrams, etc.) and another for the tables. This means that there will be Figure 1, Figure 2 . . . and Table 1, Table 2. . . .

The conventions for the numbers and titles are:

- Each table and figure in a document must have a unique number and a title.
- Table and figure numbers should be in Arabic numerals, not written out (e.g., Figure 6, *not* Figure Six) and should be assigned in the order in which the tables and figures are referred to in the text.
- In a large document the table and figure numbers can reflect the number of the section or chapter of which they are part (for example, Table 6.2 is the second table in Section 6). Try to avoid subdivisions such as Figure 6.3.2 (the second figure in Section 6.3); it becomes too complicated and is rarely necessary.
- Tables and figures in Appendixes do not belong to the two series in the main body of the document. They are labeled as two separate series in their own right, according to the numbering of the Appendix: Figure 3-2 (Figure 2 in Appendix 3); Figure C-2 (Figure 2 in Appendix C).

15.5   Titles and captions of tables and figures

Each illustration must have a figure number and a title and may have an explanation. This explanation is supplied by the caption and the legend. These two terms are sometimes distinguished from each other, but they are more often confused or considered synonymous (*The Chicago Manual of Style*, 15th ed., University of Chicago Press, 2003). To avoid confusion, this book uses only the term *caption* to describe the explanatory material that follows the title of an illustration.

- The word "Table" or "Figure" and its number are followed by two spaces and the title. There should be no period after the title unless the title is followed by a caption.
- The title must be unique and informative. It should be a phrase, not a sentence.
- The title may contain abbreviations and symbols that have been defined in the text.
- If the title runs for more than one line, the first line of the title should be the longest. The second and subsequent lines may align with the letter *T* in "Table" or *F* in "Figure" if the caption is flush with the table's/figure's left edge, or it may be indented. If the caption is centered on the figure, each line should be centered.
- If you need to identify a source, it is placed in parentheses as the last element of the caption (see Section 14.4.2).
- The table number and title are placed *above* a table; the figure number and title are placed *below* a figure. However, some graphing programs do not conform to this convention.

15.6   Conventions for tables

See Figure 15-1 for an illustration of the parts of a table.

- *Box head:* The horizontal region across the top of the table containing column headings.
- *Stub head:* The vertical column to the far left of the table in which you list the various line headings that identify the horizontal rows of data in the body of the table.
- *Spanner head:* A region that spans the head of two or more columns. Used for related parameters and to reduce repetition in the column heads.
- *Body spanner:* A region that spans across two or more columns in the body of the table.
- Column heads must all have headings. The headings should include units of measure, where appropriate, and any scaling factors used. Headings should be short; a maximum of two lines is a general rule. If

Table X. Table title

Boxhead			Spanner head	
Stub head	Column head (units)	Column head (units)	Column head (units)	Column head (units)
		Body Spanner		
Stub column	Column entry	Column entry	Column entry	Column entry
Stub column	Column entry	Column entry	Column entry	Column entry
Stub column	Column entry	Column entry	Column entry	Column entry
Stub column	Column entry	Column entry	Column entry	Column entry
		Body Spanner		
Stub column	Column entry	Column entry	Column entry	Column entry
Stub column	Column entry	Column entry	Column entry	Column entry
Stub column	Column entry	Column entry	Column entry	Column entry
Stub column	Column entry	Column entry	Column entry	Column entry
Stub column	Column entry	Column entry	Column entry	Column entry

Table note

Figure 15-1 *Parts of a table*

absolutely necessary, use abbreviations and define them in footnotes. However, avoid abbreviations if at all possible.

- Direction of reading information: The independent variable (e.g., time) usually reads across the table. The dependent variable (e.g., test number) reads vertically. Information described by the stub head reads across. Information reads down from the box head, and down from the stub head.

- Every column or spanner head needs a unit of measurement (or some explanation if the values are arbitrary rather than measurements). It is more clear to put the unit in the head rather than in the entries:

This is better...	*...than this*
Temperature (°F)	Temperature
40	40°F
60	60°F
100	100°F

- *Spanner heads* help to combine data and avoid repetition. Instead of repeating the unit of measurement after two or more column heads, a spanner head can be used:

Average daytime temperatures °F	
2004	*2005*

- To enable data to be compared: We are more accustomed to running our eyes down a column to compare data than running them across. Format the table to enable this.

- If the table is too wide to fit upright on the page, it should be presented in landscape mode so that it is read from the right-hand (outer) side of the page.
- Footnotes can be used for the following:
  — to define abbreviations
  — to explain a missing entry
  — to explain an entry that seems anomalous
  — to explain where an entry had different conditions from those in the rest of the table
  — to expand a shortened entry

However, footnotes should not take over the table. If there are too many, you need to reassess the method of presentation of the data.

See also Sections 18.7.2 and 13.8.

15.7   Formatting equations in the text

There are minor variations in styles of formatting equations. The following shows an acceptable general style.

Commentary	Text
Equation is centered.	The value of the shear stress at a distance r from the axis is given by

$$\tau = Gr\frac{d\phi}{dx} \qquad (3.5)$$

Equation number in parentheses is tabbed to the right margin. This is equation number 5 in Section 3 of the report.
In the text refer to the equation as either "Eq. (equation number)" or "equation (equation number)." Be consistent in your use of one or the other throughout your text.

Eq. (3.5) shows that the shear stress acting on the circular cross section is linear in the radius r.

▶ *For a sequence of equations in which the left-hand side is unchanged:* Align the = symbol in each line.

$$u(x) = -\frac{q_0}{AE}\int_0^x (x-\xi)\,d\xi + \frac{C_1 x}{AE}$$
$$= -\frac{q_0 x^2}{2AE} + \frac{C_1 x}{AE}$$

▶ *For continued expressions in which the left-hand side is long:* Align the = symbol with the first operator in the first line.

$$\left[(a_1 + ia_2) + (a_{11}s_1 + a_{21}s_2)\right] / \left[(b_1 + ib_2) + (b_{11}s_1 + b_{21}s_2)\right]$$
$$= f(x)g(y) + \dots$$

▶ *For expressions in which the right-hand side is long:* Align the continuing operator with the first term to the right of the = symbol.

$$V(x) = -P\langle x\rangle^0 + P(x-a)^0$$
$$+ P\langle x - (L-a)\rangle^0 - P\langle x - L\rangle^0 + C_1$$

16

Revising, Proofreading, and Reviewing a Document

All of these vital checking processes ensure the professionalism of a document and should be meticulously done. Revising and proofreading are the processes that ensure that a document is free of errors. The reviewing process ensures that it meets the Terms of Reference.

Structure of the chapter

16.1 Brief definitions: Revising, proofreading, and reviewing
16.2 Revising a document
16.3 Proofreading the final draft of a document
16.4 Proofreading the printer's proof
16.5 Reviewing a document
Checklist for revising a document

16.1   Brief definitions: Revising, proofreading, and reviewing

Revising
Revising involves deleting, altering, or adding material. The aim is to improve the technical and literary aspects of your document.

Proofreading
Proofreading involves meticulous reading and correction. The aim is to get rid of mistakes in the final draft of a document that occur in typing, type compo-

sition, preparation of figures, or in a typeset manuscript when it is returned from an editor of a journal or book immediately before publication.

Reviewing

The process of examining whether the completed document conforms to the Terms of Reference.

16.2   Revising a document

16.2.1   Guidelines: Before you start revising

▶ **Spell check the document.**

▶ **Make a printout.**

Do not try to make revisions using the monitor. The monitor does not allow you to see the document as a whole. Moreover, mistakes do not seem to be as obvious as they are on paper.

▶ **To avoid memory-reading, put the document aside for as long as possible.**

When you have read the document a number of times, you become too familiar with the text. You will tend to read from memory and miss some of the mistakes. Wait for as long a period as you can so that you approach the document fresh.

As another measure to help seeing it through fresh eyes, some people advise altering the format of a document so that it looks different from the way it appeared on the monitor.

16.2.2   The four-stage revising process

The most efficient way to revise—as recommended by professional editors—is to break down the process into four distinct components:
- Stage one: Organization only
- Stage two: Style, grammar, and punctuation
- Stage three: Formatting
- Stage four: Document integrity

Stage one: Organization only

Concentrate only on the structure of the document. Actively ignore errors of style, typography, and punctuation.

Ignoring small mistakes is not easy. When you are revising a draft for the first time, you will find that you will mostly notice errors of style, spelling, and punctuation. If you allow yourself to become immersed in this fine detail, you won't be able to pick out errors of organization, and you will have lost any advantage gained from standing back from it for a while.

While you are concentrating on the organization, rapidly make margin marks to show errors of style. Don't linger over these errors. Don't allow yourself to be drawn into correcting them. Come back to them at stage two.

The Outline mode of Microsoft Word® is invaluable for this initial stage of revising the organization of the document (see Section 3.3).

Stage two: Style, grammar, and punctuation

After revising the structure of the document, go back and polish details of style. For the individual components, see the checklist at the end of this chapter.

Stage three: Formatting

A piece of professional writing should be visually strong. It should not look boring or daunting. When trying to achieve this, make sure that each formatting decision you make is done with good reason and is consistent with other formatting; otherwise the document will become disordered. Some of the options for formatting are:

- ▶ Pack density on the page. Information is more readily absorbed if it is not too dense on the page. To achieve this:
  - Use wide margins.
  - Consider using bullets or numbering. Bullets can be a powerful method of bringing your main points to the readers' attention. But if they are used indiscriminately or too often, they can fragment a text into chaos.
  - Use indented left-hand margins, for instance, under headings, for demarcating text that you want to emphasize, or when using lists. However, avoid indenting so many times that the text is squeezed too far toward the right-hand side of the page.
- ▶ For emphasis, use **boldface type**, larger type size, or **different font** for headings. Use different fonts with discretion; the usual guideline is to use no more than two fonts in one document.
- ▶ Avoid large blocks of text set in *italics* or UPPERCASE. They are difficult to read.
- ▶ Consider justifying the text. Justified text (so that both right and left margins are straight) is used by many organizations because it looks more organized and professional. However, because of the proportional spacing, the spaces between the words can often differ markedly, which may hinder the reading process. Ragged right text is easier to read because of the evenness of the spacing.
- ▶ Avoid the following poor page breaks:
  - A heading at the bottom of the page (there should be at least two lines of text following a heading)
  - A short line (a widow) at the top of the page
  - A table that is cut in two by a page break
  - A page that ends with a hyphenated word

Stage four: Document integrity

▶ Ensure that changes made during editing do not cause discrepancies between parts of a document. Recheck the following:
  • Numbering of section headings
  • References to figures in the text
  • Correlation between text citations and the listing in the References
  • The Table of Contents and page and section numbers in the text
  • Missing figures, tables, or sections of text

▶ Use the facilities on your word processor to
  • Cross-reference figures and tables to their references in the text
  • Automatically generate a contents page (see Section 3.3)
  • Automatically correlate text citations with the citations in the References by using endnoting software

16.3   Proofreading the final draft of a document

16.3.1   Purpose of proofreading

The aim of proofreading is to rid your document of
  • Typographical errors (typos)
  • Errors that have been missed by the word processor (e.g., where you have typed "it" instead of "is," etc.)
  • Omissions
  • Irregular spacing
  • Mismatches in fonts and formatting
  • Errors in punctuation
  • Errors in spelling, particularly of nonstandard words (such as brand names) that you may have told the spell-checker to ignore

16.3.2   Importance of proofreading

Proofreading is the meticulous weeding out of errors (i.e., the capturing and correcting of detail). Even if you have created something that is well argued and structured, when proofreading is neglected it can damage your professional image and destroy your efforts to produce an impressive document.

16.3.3   Difficulties of proofreading

The major proofreading must be done after you have revised and edited for the last time. You may have done it along the way in the interests of producing something readable, but its importance lies in the final process. To proofread well, you need to be meticulous about written detail. Many people are not. Moreover, authors are sometimes poor proofreaders; because they are so close to the work, they simply do not see the mistakes. The difficulty arises from the need to abandon the usual method of reading something, which usually involves skimming and predicting what is to come. This results in seeing only what you think you wrote. The aim is to not let your document lead you by the nose and to try to read it as objectively as you can.

16.3.4   Strategies for proofreading

Many of these strategies are designed for people who usually work holistically and who do not see detail. They will force you to get away from your normal reading process, slow down, and be methodical.

- ▶ Stand back from the work for as long as possible.
- ▶ Use a printout; do not try to read it off the monitor.
- ▶ Use a pointer (a pen or pencil) or a ruler to slide down line by line. This forces you to look at each word, letter, and punctuation mark.
- ▶ Read your document aloud, either to yourself or to someone who has a duplicate copy.
- ▶ Ask a friend or colleague to proofread it as well. But do not rely on the other person too much; they will not have your vested interest in getting it absolutely correct.
- ▶ Read your draft more than once, concentrating on a different aspect each time. You could, for instance, look at the consistency of your headings—their numbering, fonts, and indentations. You will find different errors in the separate passes through your document.
- ▶ Always check paired punctuation marks such as parentheses and particularly parentheses enclosed by other brackets.
- ▶ Someone may suggest reading it backward. It is worth noting that editors do not recommend this; the process is too awkward.

16.4   Proofreading the printer's proof

The printer's proof is the typeset version that you receive back from the editor of the organization publishing your document. You may receive either galley proofs (no page numbers and with separate proofs of the figures) or page proofs (the final layout, including figures and tables, and with the final page numbers).

At this point, it is absolutely essential to meticulously proofread it yet again. You probably will have submitted the document in electronic form and may not therefore expect any errors. However, the process of setting it into its final form may have introduced mistakes, and your task is to capture and correct them.

Use the strategies in Section 16.3 yet again, but with the following additional procedures:

- ▶ Show all the corrections in the margins of the proofs, using the standard symbols (Figure 16-1). The journal may supply a list of these. If

Symbol	Meaning	Example of use
9	Delete	. . . a major factor that ~~that~~ affects the performance
∧	Insert a word *Indicate position, and write word in margin*	Two features changed during∧sampling period *the*
⤳	Insert a comma	This runoff has introduced heavy metals∧ pesticides, and pathogens.
⊙	Insert a period	Sewer overflows are a major source of pollution⊙They contribute
#	Insert a space	A rigorous∧mathematical formulation
⊃	Close up a space	A rigorous⌣mathematical formulation
⊢	Align	The following topics will be discussed: • the new management structure • the updated business plan ← • the new company logo
ital	Use italics	In the report An Analysis of Accidents in the Workplace, Brown stated . . . *ital*
Rom	Use Roman lettering (to change from italics to "standard" text)	Brown *et al* (2005) have shown that . . . *Rom*

Figure 16-1   *Proofreader's notation (continues on next page)*

not, you can find them in style manuals. The variety of symbols looks daunting; however, the ones you will use most are those for delete; insert (word, character, punctuation mark, or space); close up space; increase space; align; use italics; use Roman lettering (to change from italics to "standard" text); uppercase; and lowercase.

▶ Pay special attention to the equations. You certainly will be better able to find errors in your equations than a professional editor or proof-reader, no matter how experienced.

▶ Check the headings and the figure captions.

▶ Check the tables carefully. Complex tables can become muddled during the typesetting process. Check the vertical and horizontal align-

uc	Use uppercase instead of a lowercase letter	. . . under a sustained loading of 700 kpa
lc	Use lowercase instead of a capital letter	On the Southern side
$\checkmark$	Insert apostrophe or single quotation mark	Westergaards equations for stresses in concrete slabs
$\checkmark$	Use double quotation marks	In the report An Analysis of Accidents in the Workplace", Brown stated ...
$\sim$	Reverse these items	This report only examines two aspects.
¶	Begin new paragraph	The following topics will be discussed: (1) the new management structure (2) the updated business plan (3) the new company logo
No ¶	No paragraph	The following topics will be discussed: (1) the new management structure; No ¶ (2) the updated business plan; (3) the new company logo.
STET	Let it stand	An instruction to ignore a proofreading mark that calls for a change, and the text as originally printed to remain as it is.

Figure 16-1 *(continued from previous page)*

ment of the numerals, positioning of the headings, and numerical values against your original values.

▶ Check the final form of the illustrations. If graphs and figures have been redone, check the accuracy of the curves in graphs and the heights of bars in bar charts, the axis labels, and cropped photographs, to ensure that they have been cropped at the correct place,

▶ If the proofs are page proofs, check that the figures and tables and their captions are correctly placed and numbered.

16.5   Reviewing a document

This is a process that is usually undertaken by one or more senior managers. It is necessary to ensure that the document fully addresses the Terms of Reference. The process can be made more efficient if the completed documented is compared with an original document plan. If the plan is effective (see Section 3.2), the final document should not have drifted in any way from the original objectives.

Checklist for revising a document

Stage one: Organization
Examine your assignment critically in an overall way. Look only for errors of organization.
- ☐ Have you followed your plan for the document?
- ☐ Is the structure logical?

Stage two: Style, grammar, and punctuation

Writing for your audience
- ☐ Are you writing for your reader, not for yourself?

Punctuation
- ☐ Are the commas and periods correctly placed?
- ☐ Are the apostrophes used correctly?
- ☐ Could you use semicolons to link related sentences?

Paragraphs
- ☐ Is the first sentence an overview topic sentence?

Sentences
- ☐ Are you writing in complete sentences?
- ☐ Are your sentences too long and unstructured?
- ☐ Are they too short and disconnected?

Words
- ☐ Have you avoided apostrophes when making plurals?
- ☐ Are you using the correct word of a pair? (affect/effect, led/lead, there/their, etc.)
- ☐ Did you avoid using jargon or clichés?
- ☐ Did you avoid using pompous words where short ones would be better?
- ☐ Did you avoid worrying about split infinitives?
- ☐ Have you proofread it after the *final* spell-check?
- ☐ Did you avoid using contractions (don't, can't, etc.)?
- ☐ Are the numbers written correctly? (ten or 10)?
- ☐ Does the capitalization of headings follow the conventions?
- ☐ Have you spell-checked it?

Verbs
- ☐ Did you avoid using the distorted passive or other lifeless verbs?

☐ Are you using the right tense?
☐ Is there subject-verb agreement?

Equations
☐ Are they correctly laid out?

Stage three: Formatting

Does your document:

☐ Avoid large expanses of black text?
☐ Use listing within the text (bulleted and numbered lists) intelligently?
☐ Use emphasized text intelligently?
☐ Use no more than two different fonts (one for headings, the other for text)?
☐ Avoid large blocks of italic or uppercase text?
☐ Have effective page breaks?

Stage four: Document integrity

Illustrations

Is each of your illustrations

☐ Numbered?
☐ Titled?
☐ Adequately labeled?
☐ Correctly referred to in the text?

Text

☐ Are the headings and subheadings numbered consistently?

Table of Contents

☐ Does the wording of headings match the text headings?
☐ Is the numbering of each heading and its subheadings consistent?
☐ Is the formatting of the Table of Contents consistent?
☐ Are the page numbers correct?

Bibliographic details

☐ For each one of your text citations, is the corresponding reference in the List of References? And vice versa?
☐ Does the date of the text citation match the date in the full reference in the List of References section?
☐ Are all the references in the List of References section formatted consistently?
☐ Are all the necessary details there?

5

Writing Style

Part 5 consists of one chapter that gives guidelines for dealing with inadequate or incorrect writing style.

Engineering documents can be pompously worded or artificially inflated, and the grammar can sometimes be incorrect. Moreover, many people find it difficult to recognize whether their own writing style is inadequate. This chapter uses simple terms to deal with questions that engineers frequently ask about written style.

17

Problems of Style

Recognizing and Correcting Common Mistakes

Technical documents are often boring. This may have nothing to do with the subject matter, but rather the style in which they are written. If the style of writing is not tedious or artificially inflated, the document is less painful to read, and if the grammar and punctuation are correct, the validity of the document is greatly increased.

Many people find it difficult to recognize whether their writing has poor style. This chapter does not try to give comprehensive guidelines on stylistic elegance. Instead, using simple terms, it deals with some of the questions about style that are frequently asked by engineers.

Structure of the chapter

17.1   **Paragraphs**

17.2   **Sentences**

17.3   **Punctuation**

17.4   **Plurals**

17.5   **Pairs of words that are often confused**

17.6   **Jargon phrases to avoid**

17.7   **Writing to inform, not to impress**

17.8   **The split infinitive**

17.9   **Verbs and vivid language**

17.10 **Spell checking**

Aim when writing

The overall aim should be to inform, not to impress. Think of the clearest and most concise way to express something. The reader will be impressed not by long words and complicated constructions, but by clarity. Write as you would speak in comfortable, serious conversation, in terms with which your readers are comfortable.

17.1   Paragraphs

17.1.1   Topic sentences

In most paragraphs, the first sentence needs to be a topic sentence. This introduces the theme of the paragraph. It is needed to help orient your readers to the information you will give them in the rest of the paragraph. This is not, of course, an inflexible rule, but it is a useful yardstick. The details should then follow the scene-setting topic sentence.

> ### Example 1: Parisian sewers
>
> *First sentence (topic sentence): Gives overview of the paragraph. Remainder of the paragraph: Gives supporting details.*
>
> The history of men is reflected in the history of sewers. The sewer of Paris had been a formidable old thing. It had been a sepulchre; it had been an asylum. Crime, intelligence, social protest, freedom of conscience, thought, theft, all that human laws have prosecuted, was hidden in this pit.
>
> —Adapted from Victor Hugo, *Les Misérables*

If an engineer or scientist had written this paragraph, the order of words might have been significantly different. Our training tends to make us present the subordinate information first and draw a conclusion at the end of the paragraph, using signaling words such as *therefore*, *thus,* or *as a result*:

> ### As it could have been written by an engineer
>
> *Supporting details. Final sentence: conclusion is drawn, the rabbit is pulled out of the hat, signaled by "thus."*
>
> The sewer of Paris had been a formidable old thing. It had been a sepulchre; it had been an asylum. Crime, intelligence, social protest, freedom of conscience, thought, theft, all that human laws have prosecuted, was hidden in this pit. The history of men is thus reflected in the history of sewers.

This version does not orient the reader to the information as effectively as does Victor Hugo's version.

A method of checking the effectiveness of your topic sentences is to read only the first sentence in each paragraph over a few pages of text. If most of these sentences are true topic sentences—an overview of the theme of each paragraph—you will get a general understanding of the flow of the text across the pages. If this does not happen, check each paragraph to see whether you have instead placed the critical material at the end of it, as in the preceding example. Try bringing this material to the beginning into a topic sentence. You are likely to find that the flow of your text will be much improved.

17.1.2   Paragraph length

Do not write very long paragraphs; solid, uninterrupted text is discouraging. Varying the length of paragraphs is another way to keep readers awake.

Knowing that long paragraphs are not recommended, some writers seem to decide quite arbitrarily where paragraph breaks should be placed. The result is an incoherent text. As a general rule, use one theme per paragraph. If there is a natural break in what you are writing (try saying it to yourself), start a new paragraph.

17.2   Sentences

17.2.1   Sentence length

Short sentences are more digestible. You also can get into less trouble with their construction. Word processor grammar-checkers denote sentences of more than 25 words as too long. However, do not treat this 25-word limit as an absolute rule that should never be broken; the occasional longer sentence, if it is well constructed, not overloaded with ideas, and well punctuated, could make your writing more interesting (38 words).

Variety is important in sentence length. It keeps the reader awake. Aim for an *average* of 20 to 25 words per sentence, but oscillate around the mean.

While paragraphs should be constructed around one theme, a sentence should contain one main idea.

17.2.2   Incomplete sentences

The use of incomplete sentences—sentence fragments—is a common mistake. Without resorting to classic grammar, we will here define incomplete sentences and their rewriting.

If the main verb in a sentence is what is termed "finite," then the sentence sounds complete. Rather than define what makes a verb finite, let's just say that if a sentence seems complete, then the verb is finite:

> Many factors *affected* the resident population.
> The emissions *increased*.
> Chapter 12 *presents* the conclusions.

Now consider the following examples. In each case, the first sentence is complete; the second "sentence" is incomplete because the verb is not finite. To test this, try saying the second part of each example completely in isolation; it will seem unfinished.

> The value of our exports has dropped. The reason being the strength of the dollar.

> When running on gas, the carbon dioxide emissions were higher. Which was an indication of improved mixing and less cylinder-to-cylinder variation.

Recognizing an incomplete sentence and correcting it

There are two main ways in which incomplete sentences occur in technical writing. The first way occurs when the incomplete "sentence" has a word ending in *-ing* at or near the start of it ("The reason being" is a favorite). To correct this:

► Method 1: Join it to the previous sentence with a comma.

► Method 2: Rewrite it using a finite verb. It should sound complete in itself. Just use instinct—it usually works.

Example: Original incorrect version
The value of our exports has dropped. The reason being the strength of the dollar.

Corrected using Method 1 (a comma)
The value of our exports has dropped, the reason being the increasing strength of the dollar.

Corrected using Method 2 (using a finite verb in the second sentence)
The value of our exports has dropped. The reason is the increasing strength of the dollar.

Note: Using "The reason being is the increasing strength of the dollar" as a sentence in itself is a recently invented construction of grammatical nonsense.

Example: Original incorrect version
There are a number of strategies that countries can take. For example, promoting non-wood fuel sources, paper recycling, and pricing forest products more efficiently.

Corrected using Method 1 (a comma)
There are a number of strategies that countries can take, for example, promoting non-wood fuel sources, paper recycling, and pricing forest products more efficiently.

Corrected using Method 2 (using a finite verb in the second sentence)
There are a number of strategies that countries can take. For example, they can promote non-wood fuel sources, recycle paper, and price forest products more efficiently.

The second way incomplete sentences occur in technical writing is when the second, incomplete "sentence" starts with "which" (when it is not a question). This is very common in commercial material.

Example
All of these plans have been designed with you in mind. Which is why you'll find one that's just right for you.

However, this is unacceptable in technical writing. There are two methods for correcting it:

► Method 1: Use a comma instead of a period.

► Method 2: If using a comma makes the sentence too long, rewrite the second part. You can generally start the second sentence with "This is/was/will be…".

Example: Original incorrect version
> When the engine was running on gas, the carbon dioxide emissions were higher. Which was an indication of improved mixing and less cylinder-to-cylinder variation.

Corrected using Method 1 (a comma)
> When the engine was running on gas, the carbon dioxide emissions were higher, which was an indication of improved mixing and less cylinder-to-cylinder variation.

Corrected using Method 2 (starting another sentence using This is/was/will be...)
> When the engine was running on gas, the carbon dioxide emissions were higher. This was an indication of improved mixing and less cylinder-to-cylinder variation.

Summary: Recognizing incomplete sentences

▶ Look for words ending in "-ing" at or near the start of a sentence.

▶ Look for "which" at the start of a sentence, when it is not a question.

▶ Say it out loud. Use instinct. A sentence fragment generally will sound odd. If you say "The reason being the strength of the dollar" completely in isolation, it seems incomplete. On the other hand, "The reason is the strength of the dollar" sounds complete.

▶ If it seems incomplete, it probably needs rewriting.

17.3   Punctuation

17.3.1   The apostrophe

There are two areas where an apostrophe is used:
- The possessive—showing to whom or to what something belongs
- Contractions—where two words have been informally squashed into one

The possessive

The apostrophe shows who/what owns something.

▶ Use *'s* if there is only one owner:

the river's tributaries *(the tributaries of one river)*

▶ Use *s'* if there are more than one owner:

the rivers' tributaries *(the tributaries of more than one river)*

Note: There is no apostrophe in yours (e.g., the book is yours), hers, ours, theirs, or its (e.g., the river and its tributaries; *see also next section*).

Contractions: Don't use them in formal writing

Note: This book is not an example of formal writing, which is why we use contractions.

The apostrophe is used in a contraction to show that two words have been informally joined together. Since contractions are informal, they should not be used in the types of writing covered in this book.

The main contractions that cause confusion are:
- It's
- Who's
- Words ending in –n't (wouldn't, hadn't, etc.)

It's/its
The confusion of these is common, yet it's very easy to understand the difference. "It's" is the contracted, colloquial way to write "it is" or, less often, "it has." Therefore, never use "it's" in any formal writing. It's colloquial. It can mean only "it is."

Incorrect
> Because of overuse, the land has lost it's nutrients.

Correct
> Because of overuse, the land has lost its nutrients.

- ► Never write "it's" in formal writing. "It's" = "it is."
- ► For anything other than informal letters or notes, you will need the *its* form—with no exceptions.
- ► A good way to check: Read it aloud to yourself, saying every "it's" as "it is." Does "it is" make sense? If so, write "it is." If not, write "its."

Note: There is absolutely no such construction as *its'*.

Whose/who's
This case is very similar to it's/its. "Who's" is a colloquial form of "who is" or "who has."

Incorrect
> Mr. Smith, who's responsibility is monitoring the outfall, says that . . .

Correct
> Mr. Smith, whose responsibility is monitoring the outfall, says that . . .

Does it mean "who is"? No. Therefore, write "whose."

Incorrect
> Mr. Smith, who's responsible for monitoring the outfall, says that . . .

Correct
> Mr. Smith, who is responsible for monitoring the outfall, says that . . .

Does it mean "who is"? Yes. Therefore, write "who is."

- ► If you mean "who is" or "who has," write it.
- ► All the other times, you will need "whose."

Everything ending in ...n't (wouldn't, hadn't etc.)
In any formal writing, the words should be written out in full.

Incorrect
> shouldn't, mustn't, wouldn't, didn't, can't, hadn't, etc.

Correct
> should not, must not, would not, did not, cannot, had not, etc.

The following examples of contractions are incorrect in formal writing.

Contraction	Incorrect	Correct
couldn't	The site couldn't be examined.	could not
wouldn't	This measure wouldn't be appropriate.	would not
isn't	The river isn't polluted.	is not
wasn't/weren't	The samples weren't corroded.	was not/ were not
didn't	The water didn't contain PCBs.	did not
shouldn't	This procedure shouldn't have been followed.	should not
hadn't/haven't	The company said it hadn't been informed.	had not

▶ Never use contractions in formal writing. Write them out in full. Particularly common are:
- It's = it is
- Who's = who is
- Words ending in –n't (don't, won't, can't, couldn't, shouldn't, etc. = do not, will not, cannot, could not, should not, etc.)

Note: Plurals are not made by adding 's (see Section 17.4).

17.3.2 Commas, semicolons, and colons

Here are very brief guidelines to the main ways in which commas, semicolons, and colons are used.

Using a comma

A comma indicates a pause. You can often tell where a comma should be by saying the words to yourself. The places where commas are generally used are:

- After each item in a series, but generally not before the final *and*. The use of the serial comma is also acceptable, however, and is used throughout this book as the house style.

Example

Adjectives	The river is wide, turbulent, and muddy.
Nouns	The most common birds on the island are blackbirds, thrushes, robins, and bluebirds.
Phrases	The river-mouth is wide, with large shingle banks, extensive sand dunes, and a small island.

- To delimit a subclause from the main clause in a sentence.

Increasing agriculture will cause an increase in global warming, the reason being that ruminants and paddy fields produce methane.

When the engine was run on gas, the carbon dioxide emissions were higher, which was an indication of improved mixing.

- After an introductory phrase or subclause.

Although farmers have reduced their use of pesticides in this area in recent years, there is still local concern about the issue.

By using better management practices, farmers have been able to reduce their use of pesticides.

- To delimit material that is not essential to the meaning of the sentence.

  The island, although small, has two large river systems.

- Avoiding ambiguity when using commas (example from U.S. Navy's requirements for specifications). Commas in specifications must be used with extreme care. The presence or absence of a comma can sometimes be critical to the meaning.

Example

The flange shall be fastened by three round-head screws, three flat-head screws, and three fillister-head screws all of grade eight.

As this stands, it means that the three fillister-head screws must be grade eight. If a comma is added before "all," then you have specified nine eight-grade screws:

The flange shall be fastened by three round-head screws, three flat-head screws, and three fillister-head screws, all of grade eight.

Using a semicolon

▶ Use a semicolon between two closely related independent clauses. The statement on each side of a semicolon should be able to stand alone as a sentence.

Example

The spill caused the level of toxins in the river to rise; as a result, the entire fish population died.

▶ Use a semicolon between items in a list when the items are punctuated by commas.

Example

Yesterday I ate muesli, bacon, and eggs for breakfast; bread, cheese, and pickles for lunch; and fish and chips for dinner.

Using a colon

Use a colon to introduce a list or series.

- Before bullets

Example

The following topics will be discussed:
- The new management structure
- The updated business plan
- The new company logo

- Where the items of a list are all strung together, a colon precedes the listed points and commas or semicolons separate them:

Example

Several features changed significantly during the sampling period: water temperatures decreased; ammonium levels increased to more than 150 ppm; dissolved oxygen fluctuated; and the pH rose at one stage to 8.3.

17.4   Plurals

Never make a plural—more than one of something—by adding *'s*. Just add *s*. There are only two types of exceptions to this:

▶ Add –es if the word ends in –o or –sh (*potatoes or fishes*)
▶ Add –ies if the word ends in –y (*ferries*).

The plurals of abbreviations or dates do not have apostrophes.

Incorrect	*Correct*
Many component's	Many components
Many potato's	Many potatoes
Many fish's	Many fishes
Many pizza's	Many pizzas
Many ferry's	Many ferries
PCB's	PCBs
The 1990's	The 1990s

17.5   Pairs of words that are often confused

There are pairs of words or expressions that are often mixed up. Some of the most common pairs are:

• Affect/effect
• Compliment/complement
• It is composed of/it comprises
• Its/it's
• Lead/led
• Loose/lose
• Passed/past
• Principal/principle
• Their/there
• Which/that

17.5.1   Affect/effect

Focusing on the most common uses of the two words in most technical writing: "affect" is a verb, "effect" is a noun (see the Quick Reference Guide: Parts of Speech and Verb Forms in Part 7 for guidelines on these terms).

• To **affect** something is to influence it (*a verb*).

   The pollution will affect the dissolved oxygen concentration.

   The pollution has affected the dissolved oxygen concentration.

• The **effect** of something is the result or consequence of it (*a noun*)

   The pollution will have an effect on the dissolved oxygen concentration.

Incorrect

Natural events such as volcanoes or eruptions may effect the ozone layer.
The fish numbers were severely reduced, thus effecting the food chain.
This report examines the affects of improving this system.

Correct

Natural events such as volcanoes or eruptions may affect the ozone layer.
The fish numbers were severely reduced, thus affecting the food chain.
This report examines the effects of improving this system.

Rule of thumb for the use of affect/effect

▶ Use "affect" as a verb: This affects something, this will affect something, this has affected something.

Correct

This measure will affect the waterway.

Incorrect

The affects of this measure will be permanent.

▶ In most cases, use "effect" or "effects" as nouns. This means *a, an, the* effect or *the, some, several, a few, many, a couple of* effects.

Correct

The effects of this measure will be permanent.

Incorrect

This will effect the waterway.

Note: "Effect" can also be used as a verb, but it has the very specific meaning of causing something to happen.

Example

This measure will effect a significant change.

17.5.2 Compliment *and* complimentary/Complement *and* complementary

▶ Use "compliment/complimentary" where you want to imply flattery.

Examples

He complimented the guest speaker on her presentation.

You will receive a complimentary ticket to the dinner.

▶ Use "complement/complementary" to imply the fitting together or completing of something, or the finishing touches. "Complementary" is also used in technical or mathematical terms in the sense of fitting together.

Correct

The landscaping at the front of the building complements its radical form.

Incorrect

The landscaping at the front of the building compliments its radical form.

Also

Complementary angles; complementary color; complementary relationship; complementary function

17.5.3 Composed/comprises

There are two expressions that are commonly confused: "is composed of" and "comprises."

Incorrect

It is comprised of three elements.

Correct

It is composed of three elements.
It comprises three elements.

Rule of thumb

"Comprised," "comprises," and "comprising" can never be followed by "of."

17.5.4 Lead/led

This has become confused because "lead" is pronounced in two different ways: (1) as the element Lead (Pb) and (2) as in "He will lead the team." (This word does not follow the same system as "read," which is what often confuses people.) The most common misuse is:

Incorrect

This has lead to pollution of the stream.

Correct

This has led to pollution of the stream.

A rule of thumb for lead/led

Whenever you write "lead," say it to yourself. Does it sound like "led"? If so, you mean the element Lead (Pb). Does it sound like "leed"? If so, write "lead."

17.5.5 Loose/lose

"Loose" is often used incorrectly, where "lose" would be the correct form. In their most usual senses in technical writing, "lose" means "to cease to possess" or "misplace"; "loose" means "not restrained," as in "a loose fit" or "the cover was loose."

Incorrect

The breeding pairs will loose their chicks if conditions do not improve.

Correct

The breeding pairs will lose their chicks if conditions do not improve.

17.5.6   Passed/past

These two terms are best explained by examples of their use:

- Passed

  The law that has just been passed states that ...
  This material was not passed to the other team members.

- Past

  In the past...
  Past practices...
  Over the past year...
  The road runs past the waterfall.

17.5.7   Principal/principle

These two terms are best explained by examples of their use:

- "Principal": the most important, the highest in rank, the foremost

  The study was made up of five principal sections.
  The principal of the institution said that...

- "Principle": a fundamental basis of something

  Archimedes' principle...
  The chief investigator had no principles.
  The principles of the investigation were...

17.5.8   Their/there

A rule of thumb for their/there

- "Their" is never followed by a verb (such as is, was, will, can, should, would, could, may, might).

  Incorrect
  Their was; their is; their could be/should be/would be; their will; etc.

  Correct
  There was; there is; there could be/should be/would be; there will; etc.

- On all other occasions (except when you are saying something is "over there"), you are likely to need "their":

  Incorrect
  Their are a number of steps that can be taken.
  There advanced features make them superior to the previous models.

  Correct
  There are a number of steps that can be taken.
  Their advanced features make them superior to the previous models.

17.5.9   Which/that

The use of either "which" or "that" can alter the meaning of a sentence.

> *Example using "which"*
> The report, which was released on Monday, showed that the level of sewage contamination had decreased.

The commas around the clause show that it can be left out without affecting the meaning of the sentence. That is, it may be of interest to know that the report was released on Monday, but it is not essential.

> *Example using "that"*
> The report that was released on Monday showed that the level of sewage contamination had decreased.

The clause "that was released on Monday" cannot be omitted without changing the meaning. In this form, the sentence means that a particular report (the one released on Monday) showed that contamination had decreased. Another report released on Thursday, for instance, could have shown the opposite.

This is ambiguous:

> The report which was released on Monday showed that the level of sewage contamination had decreased.

17.6   Jargon phrases to avoid

Prepacked, clichéd jargon phrases have no place in formal writing. Some common phrases to avoid are:

a window of opportunity	in the long run
all things being equal	in the matter of
as a last resort	it stands to reason
as a matter of fact	last but not least
at the end of the day	level playing field
at this point in time	many and diverse
comparing apples with oranges	needless to say
conspicuous by its absence	on the right track
easier said than done	par for the course
effective and efficient	slowly but surely
if and when	the bottom line
in the foreseeable future	

17.7   Writing to inform, not to impress

English has a number of pairs of words that have equivalent meanings. The longer words are of French origin, and, for formal writing, are almost always chosen in preference to the corresponding short words, which are of Anglo-Saxon origin. The two underlying reasons for this are: (1) writers often believe that the longer words make the report more impressive, and (2) the French-origin words have come to have a softer meaning than the Anglo-Saxon; writers therefore use them to hide behind their meaning and avoid the bluntness of the shorter word.

Word of French origin	*Word of Anglo-Saxon origin*
anticipate	expect
assist	help
commence	start
desire	want
endeavor	try
indicate, reveal	show
locate	find
purchase	buy
request	ask
require	need
terminate	end
utilize	use

However, writing that consistently uses the longer words can be very pompous. A useful guideline is: don't avoid using the longer words altogether. Avoid using them exclusively, and aim for a mixture of long and short, using the shorter word whenever it does not look too blunt. This will help your reader to avoid boredom.

Example of inflated style
In our endeavor to ascertain whether or not the proposals we have formulated are sound, we anticipate engaging a survey organization to assist us in determining whether our conception of the market is substantiated by information accumulated in the field. (*40 long words and a complex construction*)

Rewritten
To find out if our proposals are sound, we expect to hire a survey group to help us check our view of the market against facts gathered in the field. (*30 shorter words and a simpler construction*)

17.8   The split infinitive

What is an infinitive? It is a verb form, the first word of which is "to" (to construct, to prepare, to analyze, etc.). A *split infinitive* occurs when words are placed between "to" and the other word.

Example
They were urged <u>to seriously reconsider</u> their stand.

Split infinitives are not nearly as important as they are often made out to be. There is not, as many people suppose, a rule that says an infinitive should not be split; it is merely an invention observed by them in the mistaken belief that they are showing their knowledge of "good" writing. There is no valid basis for this outmoded idea.

Pedants would insist that the sentence above should be rewritten. However, the three rewritten forms sound awkward:

> They were urged to reconsider seriously their stand.
>
> They were urged seriously to reconsider their stand. *(This is ambiguous.)*
>
> They were urged to reconsider their stand seriously.

Sometimes a split infinitive is needed to avoid ambiguity:

> He would like to really learn the language.

The alternatives are ambiguous: "He would like really to learn the language" could mean the same as "He would really like to learn the language."

However, a lengthy interruption in an infinitive is not good style:

> The political will is lacking to resolutely, wholeheartedly, and confidently reform the tax system.

When considering infinitives, write whatever sounds the least awkward. There is no valid grammatical reason to avoid split infinitives. However, you need to be aware that many people still see them as an indication of poor writing, and many newspapers still avoid them. So avoid writing them if possible.

17.9   Verbs and vivid language

Vivid language is not something that most people associate with technical writing. Yet if readers are given dull, impersonal prose, they become bored. A lot of technical writing is dull, and some of the problem has to do with the way we use verbs.

17.9.1   Active or passive voice?

Many writing handbooks and word processor grammar-checkers tell us to use the active voice of the verb, not the passive. This is not useful advice—many people do not know what the active and the passive voices of the verb are. Even textbooks are sometimes confusing; there is a lot of advice to write actively and therefore use the active voice. This is nonsense. The words *active* and *passive* are the classic names for the voice of the verb; despite their imagery, they are not related to the liveliness of the prose.

We will now ask the following three questions:
1. What is meant by the active and passive voices?
2. Is it poor style to use the passive voice?
3. What happens when we distort the passive voice and make *really* pompous sentences?

1. What is meant by the active and passive voices?

The following sentence has a subject (or actor), "acid-etching"; a verb, "removed"; and an object or receiver, "rust."

> *Active voice of the verb*
> Acid-etching removed the rust.
>   *actor*   *verb*   *receiver*

This sentence is in the active voice because the order of the flow is actor, verb, receiver. If this sentence is turned around, we have:

> *Passive voice of the verb*
> The rust was removed by acid-etching.
> *receiver*   *verb*   *actor*

When the order of flow is receiver, verb, actor, the sentence is in the passive voice.

When an active sentence is turned around into the passive voice:

- The emphasis changes. In the active sentence, the emphasis was on "acid-etching"; in the passive form, "rust" is emphasized.
- The order of the flow is reversed.
- The number of words in the verb increases—"removed" becomes "was removed"—as a result of adding forms of the verb "to be."
- An extra word is needed ("by").

2. Is it poor style to use the passive voice?

No. The passive voice is not intrinsically poor, despite what many writing textbooks and grammar-checkers tell us. We need the passive voice; it stops us from repeatedly having to use "I" and "we" or some other agent. In technical writing, we would write, quite naturally:

> The site was inspected last month. (*passive*)

implying

> The site was inspected by someone last month.

In the active version, we would need to specify who examined it

> *I/we/the contractor/Mr. Brown/someone* inspected the site last month.

This is normally not necessary. In technical documents, it is usually what was done rather than who did it that is important. Moreover, if "I" or "We" are used too often, it can sound like a child's description of a day out.

Often we can actively choose which voice of the verb to use. For instance, if we were writing a paragraph about bees and their relationship with pollen, we would write:

> *Active*
> Bees carry pollen.

If the paragraph were about pollen, we would write:

Passive
> Pollen is carried by bees.

Each of these is completely acceptable; it depends on the emphasis that we need.

3. What happens when we distort the passive voice and make *really* pompous sentences?

What *is* bad is to take the passive voice one step further into a distorted form. Then the verb becomes hidden in a sort of noun, and pompous, predictable main verbs have to be used. This happens often in technological and scientific writing.

Let's consider the progression in these sentences:

> Acid-etching removed the rust. *(active voice, acceptable)*

Turn this around and it becomes:

> The rust was removed by acid-etching. *(passive voice, acceptable)*

If the verb "was removed" becomes hidden in a sort of noun placed earlier in the sentence, it becomes:

> Removal *(hidden verb)* of the rust was … *(missing verb)* by acid etching *(distorted passive, tedious and pompous)*

Ask someone to insert the missing verb and the suggestions are always the same. The favorites are: **achieved, accomplished, carried out, effected, performed, undertaken.**

Now we have lost the skeleton of the sentence. We have gone from "Acid-etching removed," or "The rust was removed"—both of which are good—to "Removal was carried out," "Removal was achieved," and so on, which sound pompous.

This distortion is a common way of writing tedious prose in formal engineering writing. It often sounds completely normal, because we have become accustomed to seeing it:

Examples
> The ohmmeter measured the resistance. *(active voice, acceptable)*
> Resistance was measured by the ohmmeter. *(passive voice, acceptable)*
> Measurement of the resistance was carried out by the ohmmeter. *(distorted passive, unnecessarily inflated, pompous)*
>
> I examined the site daily. *(active voice, possibly acceptable)*
> The site was examined daily. *(passive voice, the acceptable style for technical writing)*
> Daily examinations of the site were carried out. *(distorted passive, unnecessary distortion)*

A method of seeing its absurdity: We are so accustomed to seeing the distorted passive in professional writing that the absurdity of the construction is obvious only when it is seen in an everyday context:

Hansel said, "The birds have eaten the crumbs." *(active, acceptable)*
Hansel said, "The crumbs have been eaten by the birds." *(passive, acceptable)*
Hansel said, "Eating of the crumbs has been carried out by the birds." *(distorted passive, absurd in this context)*

Summary: Rewriting the distorted passive

▶ If you find yourself using words such as **achieved, accomplished, carried out, effected, performed,** or **undertaken,** you are very likely to be in the distorted passive. So keep these words in mind as danger signals.

▶ Find the hidden verb. It will be earlier in the sentence, probably in a word ending with *–ing, –tion,* or *–ment.*

▶ Use it to rewrite the sentence, using either a simple passive construction or the active.

17.9.2   Excessive use of nouns instead of verbs

Some lifeless verbs can mutate into nouns, and the pace slows down:
- "Indicates" becomes "is an indication of."
- "Suppose" becomes "make the supposition."

Original lifeless version with verb mutated to noun
The color of the outfall was an indication of severe pollution.
We may therefore make the supposition that ...

Rewritten using verb
The color of the outfall indicated (or showed) severe pollution.
We may therefore suppose that ...

17.9.3   Subject-verb agreement

Make sure that the subject (actor) of your sentence agrees with the verb.

Original incorrect version
A vast network of suburbs stretch out from the center.

Corrected version
A vast network of suburbs stretches out from the center.

Note: The verb is referring to "network" (singular), not "suburbs" (plural).

Original incorrect version
The greatest loss of lives as a result of a volcanic eruption have occurred through pyroclastic flows and tsunamis.

Corrected version
The greatest loss of lives as a result of a volcanic eruption has occurred through pyroclastic flows and tsunamis.

Note: The verb is referring to "loss" (singular), not "lives" (plural).

17.9.4   The correct tense/form of the verb

For simple examples of forms of the verb, see the Quick Reference Guide.

Decisions about the proper use of tense can be confusing. There are no absolute guidelines. Here are suggestions that follow the general conventions for deciding which tense to use in technical documentation:

Use past tense for describing:

- Procedures and techniques

  Tailing soils were treated by presieving. (*You are describing a procedure.*)

- Results (yours and other people's)

  The cation-exchange capacity of the sediments was low (Brown, 2000). *(You are describing other people's results.)*

Use present tense for describing:

- Established knowledge, existing situations, and the theoretical background to a study

  Rubber subjected to a hydrostatic tensile stress causes *(present tense)* internal rupture known as cavitation.

- Your conclusions

  It may be concluded that the method accurately predicts (*present tense*) flow parameters of the subcritical flow.

- Illustrations in the text

  Figure 3 shows (*present tense*) the effect of temperature on the solubility of the salt.

- Geologic and geographic features

  All three paleosols show (*present tense*) a greater degree of development than the surface soils. Better development is (*present tense*) displayed in terms of greater clay accumulation, higher structural grade, harder consistency, and thicker profiles.

Specific uses

- The conditional, subjunctive, or imperative forms can be used when giving recommendations (see Section 5.2.6).
- The imperative form is used in procedures or sets of instructions (see Section 8.4).

Summary: Verb tense

As a general indication:

- ▶ Use past tense for almost everything, but use the present tense for describing established knowledge, existing conditions, and theory; illustrations; and geologic and geographic features.
- ▶ Use the conditional, subjunctive, or imperative forms for Recommendations.
- ▶ Use the imperative form for Procedures.

17.10   Spell checking

All documents should be run through the spell checker before submission. However, there are two main problems associated with this:

- In the final rush, it is easy to insert additional information after what was thought to be the final spell check. Make sure that the very last stage of a document's production life is a spell check.

- It is still possible to produce meaningless prose. The spell checker will pass "as" where you meant "at," "it" where you meant "is," and so on. A document should always be meticulously proofread after the final spell check. If you know you are poor at this, give it to someone else.

Presenting Work Orally

Part 6 deals with the professionally necessary but often anguished process of oral presentation of material. The ability to be able to present technical material in different situations is a critical competency for a professional engineer. Moreover, many people are concerned about their performance and level of nervousness. The material here is the result of having watched presentations of hundreds of students making their first tentative steps into the process, and those of many experienced professional engineers and conference participants. They often all make the same mistakes.

Guidelines are given both for formal presentations to a large audience and for those to small groups such as an oral examination, a professional interview, or a review panel.

18

A Seminar or Conference Presentation

Presenting work orally to an audience can be unsettling or nerve-wracking for many people. The problem can have an additional dimension if you are a nonnative speaker in an English-speaking venue.

Those who advise on public speaking often appear to forget that much of it cannot be put into practice by a beginner or nonnative speaker. Many people are worried only about surviving the experience with their credibility more or less intact; they do not want to think about the subtleties of interacting with an audience and other sophistications.

During programs that I have run, hundreds of engineers—from freshmen to senior company personnel—have given an initial, prepared presentation and then had a second attempt, following the guidelines given in the feedback discussion sessions. The second presentations were, in every case, markedly improved over the first. The main reason for the improvement has always been that the structure of the material was much better, with readily recognizable main points. A presentation will be judged on whether the audience can readily access and therefore remember the main information. Without a good structure, no amount of colored graphics and speaking skills will make a professional presentation.

This chapter describes the characteristics of effective professional presentations of technical material and gives guidelines for both beginners and experienced people. It shows how to develop a good structure for your material, with visual aids that reflect it. It also describes common mistakes; an awareness of the things that people can do when they are nervous can help you avoid or minimize them.

This chapter does not give guidelines on how to produce effective presentation graphics. Specialized graphics books should be consulted for this.

Structure of the chapter

18.1   The aims of a presentation and the constraining factors

- To present your work orally to an audience that can be expert, nonexpert, or mixed
- To be able to adapt—often in a strange room—to the equipment provided for visual aids
- To be able to present within a fixed time limit that beforehand can seem very long, but is often too short

18.2   Guidelines for beginners

Accept that you probably are going to be nervous.
Plan strategies to deal with it.
Since the thought of being nervous is a major concern for many people, we will deal with it here before we go on to the other aspects.

The symptoms of nervousness that are relevant to making an oral presentation are thumping hearts, shaking legs and hands, rapid breathing, and voices that crack embarrassingly. These can sometimes hit unexpectedly, just when the chairperson is introducing you. Here are a few ideas on what can be done about it.

> ► Make sure there is water for you to drink immediately before you get up to speak, as well as on the podium. Sipping water can have a calming effect.
>
> ► Deep breathing can calm some people, but not everyone. Try it and see. Breathe from the diaphragm so that your belly moves, not from the chest; chest-breathing and gasping can make you giddy.
>
> ► Walking briskly to the lecture theater is sometimes suggested. It could

work. Try it and see, but don't run; running increases your need for oxygen and can result in an even greater heart and breathing rate.

▶ The only effective remedy: accept that when you get up to speak, you may be nervous, but you know that you have strategies to deal with it.

The most effective strategy for dealing with initial nervousness: Know the first 30 seconds or so of your talk—your introductory material—by heart.

When you are nervous, you are often on automatic pilot. Your mind can close down, and you may not be aware of other things at all. This state usually improves after a minute or so. But during this time, you need to be sure that you can deliver the first part of your talk. Practice it.

After your first few words of greeting, show your Title slide. Then follow with your Overview slide.

The first two slides of any presentation should always be a Title slide, followed by an Overview slide. The Title slide has the immediate effect of moving your audience's gaze away from you and onto the screen. It's a relief. It also has the effect of tuning in the audience to you and your work.

The Overview slide gives you something solid and undetailed to start with. This has a calming effect. It is also essential for the audience's understanding of the later material.

Don't worry about a cracking or squeaking voice.

You may feel that your voice is cracking embarrassingly. However, it is very common to find that your colleagues afterward will tell you that it was unnoticeable. Remember that what sounds to you inside your head like a peculiar voice may not sound very different from your normal voice to an audience. Don't worry about it.

Try not to read.

It is very boring for the audience. But if you do need notes, format them for easy reading and pathfinding.

If you must resort to reading your script, try to look up at the audience as much as possible.

It gives a poor impression if you never look at the audience. Try to force your gaze up as much as you can, even if it's all a blur. It is a common mistake to ignore the audience by taking refuge in looking at the screen or the notes too much. You can also trick the audience into thinking you are looking at them (see Section 18.8.5).

Longer term strategy: Take every opportunity to give talks.

It's a tough strategy, but it's only by frequent practice that nervousness is eventually conquered. Even very good presenters have suffered from nervousness when they were beginners.

Be aware of the things that people can do when they are nervous.

The mistakes listed in this chapter are not meant to be depressing. Being warned of them beforehand can help you develop strategies to avoid or minimize them.

18.3   Structuring the presentation

Overview

- ▶ Make sure that the presentation deals with only a few main points and is well structured.
- ▶ Make sure you go through the following procedure: greet the audience; show title slide; show overview slide; give context of the work; provide detail; show conclusions slide; thank audience and take leave.

18.3.1   Overall points to remember

- How the talk is structured is vital to a listener's understanding.
- Anyone in the audience, whatever his or her level of knowledge, should be able to understand the broad concepts of what you are talking about. There is no complex system, no computational method—nothing—that cannot be explained simply enough for this to happen, provided the structure of the material in the presentation is effective.
- It is a far greater achievement to be able to express a complex idea clearly than to bombard the audience with detail.

18.3.2   The structural differences between an oral presentation and a written paper

When people absorb and process facts, there are large differences between the way they do so when they read the material and when they listen to it. This has implications for structuring an oral presentation.

Table 18-1 compares how technical material is understood and assessed by a reader and by a listener. The processes are both quite different, and this should be taken into account when you prepare a talk.

For these reasons, an oral presentation needs a type of structure and wording to help the audience engage with the material. The manner of giving it will also play a large role.

18.3.3   Steps in planning a presentation

First step

Decide on the information that you most want the audience to remember about your talk. Make sure that this will be clearly stated three times: in the initial overview slide, in the main body, and in the concluding slide.

Second step

Decide on the rest of the information. You will be able to cover far less than you would like, so you need to be very selective.

Table 18-1. *Comparison: Assessment of technical information when reading a paper and when listening to an oral presentation*

When reading	When listening to an oral presentation
Can be read at the reader's own speed.	Listener has no control over speed. Understanding is impaired if the material is presented densely or at speed.
Can be reread.	No opportunity to increase understanding by a rerun.
Overall understanding gained beforehand from the overview information in the Abstract.	Listener needs an initial overview.
Headings, subheadings, diagrams can be scanned beforehand.	No opportunity to scan ahead.
Material can be skipped.	If listener switches off, he or she may not readily switch on again.
Technical writing is neutral—it does not usually convey the writer's enthusiasm or lack of it.	Your manner of speaking can convey your enthusiasm or boredom with your work.

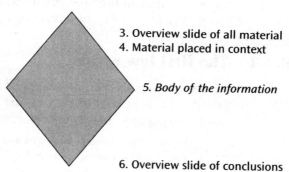

1. Greeting; introduction of yourself

2. Title slide

3. Overview slide of all material
4. Material placed in context

5. *Body of the information*

6. Overview slide of conclusions

7. Take leave

Figure 18-1 *The presentation structured as a diamond of detail: start with low detail (Overview slide); main body of information (detailed); finish with concluding overview (Conclusions slide)*

Third step

Structure the talk as a diamond of detail (Figure 18-1). This scheme ensures the following:

- It provides an initial overview of all the material, which is vital for your listeners to be able to assess the subsequent information. It also greatly helps the presenter.

- It provides a summing up of the conclusions at the end, which confirms the main points of your talk in the listeners' minds. This is also useful if you have to finish in a hurry (see Section 18.8.10).
- It ensures that your take-home messages—your main points—are heard three times: (1) in the initial overview, (2) in the main body, and (3) in the concluding section. This follows a well-known principle for presenting information:
  — Tell them what you're going to tell them.
  — Tell them.
  — Tell them what you've told them.

The activities of the diamond-structured presentation are shown in Table 18-2, together with the mistakes that are commonly made at each stage.

18.4 Suggestions for wording: Your own, and for visual aids

Overview: Wording
- The first few words: greeting the audience; introducing yourself; the title visual aid
- Wording of the second visual aid: the Overview
- Wording of the final visual aid: the Conclusions
- The final few words: finishing up

18.4.1 The first few words

At a conference, even if the chairperson has just introduced you by name and stated the title of your talk, don't just plunge into your material. Go through the following sequence. Greet the audience by saying (approximate wording):

> Good morning/afternoon, ladies and gentlemen. (*Look at them, even if you're nervous.*) I am/My name is (*your name*); I'd like to present the work we have been doing on...

This orients your audience and is a more polished way than merely repeating verbatim what the chairperson has probably just said in his or her introduction of you.

During the greeting, put up your first visual aid: the Title slide.

18.4.2 The first visual aid: The Title slide

Immediately after you have said good morning and while you are saying where you work and what you have been doing, put up your title visual aid. This should state the title of your talk, your name, and your institution. The audience's gaze will move away from you and onto the screen, and you will

Table 18-2. *The activities of the diamond-structured presentation and common mistakes*

Activity	Common mistakes
Greet the audience: Look straight at the audience and introduce yourself.	Forgetting to greet. Unprofessional greeting and shuffle. Forgetting to look up because of nervousness.
Title slide: Say what you are going to talk about.	Title slide often omitted or not seen as necessary.
Introduction: Overview slide. Summarize the material in point form. Make sure you give real information: informative, not descriptive (see Chapter 6). This helps the audience to assess the information that will follow.	Very often omitted, because significance to audience and yourself is not appreciated.
Give the context of your material—the overall view of the field and where your material fits in.	No big picture, only a mass of detail. Audience does not engage.
Present the body of the information. Make sure there is a clear, well-constructed framework and that it is obvious to the audience (possibly: method of approach; results; implications).	Mass of facts, no framework evident: audience therefore loses track.
The final few minutes: A conclusions overview slide giving the conclusions from your material. Important section: It clearly presents the significance of your work and reinforces your material in the listeners. It is a far more effective way to finish your talk than merely coming to an unstructured end. It is also vital if you need to finish in a hurry.	Very often omitted, because significance to audience and yourself is not appreciated. Unprofessional ending. Presentation abruptly stops or trickles to a close.
Take leave of your audience in very few words. The most professional way: Look straight at the audience, perhaps give slight a nod and smile, and say "Thank you."	Unprofessional shuffle and wording. *Unprofessional:* "Er, well, that's about it" or similar. *Too artificial:* "I'd like to thank you for your kind attention in listening to this talk," or similar.
At a formal presentation: Stop. Remain standing. Don't ask for questions.	Asking for questions (this is the chairperson's task).

feel less exposed. The audience needs this too; it gives them immediate access to the main feature of your work, together with details about yourself.

18.4.3   The second visual aid: The Overview

This can be called Overview or Summary. Briefly list the main points that you are going to cover, and then fill them in with the spoken material. This gives the audience the initial brief orienting overview.

A. Effective list of overview points.

Note: You don't need to say all of these points while dealing with this slide.

Roof design of new soccer stadium, Cheju, S. Korea

Overview

Location	In area of outstanding natural beauty
	Reflects cultural heritage: local fishing boats, volcanic mountains, crescent moon
Characteristics	Soaring, crescent-shaped fabric roof
Challenges	• Unusual geometry
	• Complex structural system
	• Substantial wind uplift forces
	• Unique shape, lightweight materials: wind load needed accurate determination

Wind tunnel testing to determine accurate force coefficient

Optimized structural design of much greater concern than usual

Aggressive construction schedule

B. Ineffective list of overview points

These are especially meaningless if read out verbatim.

Roof design of new soccer stadium, Cheju, S. Korea

Overview

Location

Characteristics

Challenges

Methods of testing

Optimization

Conclusions

Figure 18-2 *Effective and ineffective overview slides*

Content adapted from: Lee, J. C. (2002). "Arc of triumph." *Civil Engineering*, August, 38–47.

Aim for real information (avoid meaningless headings) and for concepts, words, and phrases that will engage the interest of the audience (as in Figure 18-2).

Common mistake

- Avoid using a series of uninformative headings and saying something meaningless along the lines of

  "I'm first going to introduce the topic, then speak about the location and characteristics of the roof; I'll then describe the challenges we faced and our methods of testing, and finally give you our conclusions."

However, even this is better than no overview slide at all.

**Roof design of new soccer stadium,
Cheju, S. Korea**

Conclusions

Highly successful because:

- Completed to schedule
- Final cost $90 M less than was budgeted
- Combined ingenuity of South Korean and American firms
- Environmentally sensitive, visually engaging, draws on cultural images

Quotes:

"Arguably the most spectacular football stadium of the twenty venues built or redeveloped for the World Cup."

"Its setting is one of the most beautiful on the planet."

Figure 18-3 *An effective Conclusions slide*

18.4.4　The final visual aid: The Conclusions

This should be titled Conclusions or Concluding Overview and should be a list of the conclusions you can draw from the work that you have presented. Use this slide for the last main slide of your talk. This reinforces your work in the minds of your audience and is a very effective way to close your talk.

Figure 18-3 shows a concise list of major conclusions, together with a set of quotes with which to finish the presentation.

Use appropriate wording while you put up the Conclusions slide, e.g., "So, finally, I'd like to summarize the main points of what we can conclude from this work." Then briefly run through the list. Try to avoid reading it word for word.

Common mistake

Avoid displaying the same or a similar list as in the initial Overview visual aid. The two should be quite different. The first is an overview of the whole talk; the final one is a list of the conclusions—what you think it all means.

18.4.5　The final few words: Finishing up

The best way to take your leave is to look at the audience, say a firm thank you with perhaps a slight nod of the head and a smile, and stop. Remain standing while the chairperson asks for questions.

Common mistakes

- To end very feebly by a body shuffle and words such as "Well, that's all I've got to say, really," or "That's it—so-um—thank you."

- Anything along the lines of "I'd like to thank you for your kind attention in listening to this paper." It sounds inflated and artificial.
- Asking for questions. In a formal conference or seminar, it is the role of the chairperson to ask for questions and to choose the questioners. Just conclude, keep standing there, and wait for him or her to take over. (For guidelines on answering questions, see Section 18.9.)

18.5   Types of speaker's notes

Overview: Types of speaker's notes
- Brief notes
- Cards
- An annotated full script

Most people are so familiar with their topic that they do not need notes except as a form of reassurance. The danger of this is that when notes are present, they will almost inevitably be read. It is important to be aware of what can happen when a script is read.
- Your voice could become monotonous and dull.
- You could lose all appearance of enthusiasm and bore the audience.
- Your stance could become rigid.
- The audience could feel that you aren't in contact with them, since you aren't looking at them.
- You could lose your place when you have to take your eyes off your script to deal with your visual aids. Then you have the embarrassment of trying to find your way back into it.

The best strategy for avoiding notes is to take all your cues from well-constructed points on the monitor, overhead projector, or screen. But if you think this is impossible, here are some suggestions.

Brief notes or Cards
Both are for the relatively confident. They are much better than a full script, because they do not allow you to read word for word.

Format them as a series of headings. Use large, lowercase type; uppercase type can be difficult to read. It may be useful to include the phrases needed to change direction and to mark the points where you need to change the visual aids. Number them in case you drop them.

Annotated full script
Despite much advice to look at the audience and not to read, many beginners often prepare a full script, intending to use it just for security and not to read it. However, they often find that their nervousness makes them resort to reading it. An alternative is to use an annotated full script. This enables you to work from keywords, but it also gives you the reassurance of the full script if you need it.

- Divide your legal-sized paper so that one-third of the sheet becomes a very large left-hand margin.
- Write your full text in the remaining two-thirds. Remember to use spoken English, not written (see Section 18.6) and to use a large type size so that you can see it easily. The venue may have dim lighting.
- In the wide margin, alongside each part of your complete script, write keywords that are relevant to the text opposite. Use large letters so they are easily read. Identify them (e.g., by color) so that you can readily distinguish among the various points.
- Number the sheets in case you drop them.
- During the talk, if you do not need to read the script, run vertically down the left-hand margin, using the keywords to prompt you.
- If you need to read at a particular point (e.g., if you need to quote accurate data), write *READ* in the margin.
- As you finish each page, lift it off and lay it down alongside, keeping the pages in order. A few words of warning:
  — Don't staple the sheets together: They are difficult to manipulate and can rustle.
  — Don't use double-sided copies: You can easily lose your place and a double-sided stapled script is a disaster.
- If you lose your place, stay quiet, don't fluster, control your body language, and navigate your way back by using the keywords. If you don't make it obvious, no one will notice.

18.6   Spoken style

Overview: Spoken style
- ▶ Use spoken English, not written English.
- ▶ Use simple clear words, but include the correct technical vocabulary.
- ▶ Don't be afraid of using the personal *I* or *We*.
- ▶ Don't read out subheadings.

Use spoken English, not written English
The whole point of an oral presentation is to *talk* to your audience. If you are going to read the script, you might as well give it to someone else to read.

If you are nervous, you may find yourself relying on your notes more than you expected. But if your words sound as though you have thought them out as a spoken presentation and not as a written one, you have a much better chance of not boring your audience. Written English sounds flat and dull. Enthusiasm is infectious, but so is a lack of it.

Use simple, clear words, but include the correct technical vocabulary
Think in terms of the style of comfortable, serious conversation. To do this, imagine yourself explaining your work across a table to a colleague, comfortably and without using colloquialisms.

Don't be afraid to use the personal We or I

If it is not used too often, it livens up a presentation.

Highly appropriate (spoken style):
We looked at the literature to establish whether . . .

Inappropriate and boring (inflated written style, probably read from a script):
A literature survey was undertaken in order to establish whether . . .

Don't read out subheadings

This shows that you have devised your talk in written terms instead of spoken terms.

Common mistake

From novice speakers, it is common to hear oddities such as:

Design objectives. The design objectives were . . .

Sampling methods. Three sampling methods were used. . .

18.7   Designing visual aids

Overview: Designing visual aids

► Reinforce what you are saying by simultaneously showing it in abbreviated form on the screen.
► Make sure that the audience can read the text and see the diagrams clearly.
► Use an uncluttered layout.
► Use point form: Do not write complete sentences.
► Proofread your visual aids for spelling.
► Be aware of how to use presentation software effectively.

18.7.1   Reinforce what you are saying by simultaneously showing it in abbreviated form on the screen

Remember, an audience will absorb and recall information far more efficiently if they hear and see it simultaneously.

It is not enough for your visual aids to show only diagrams and illustrations. You also need to have informative text on the screen in abbreviated form that reflects what you are saying at any time. Your spoken words should expand on the information on the screen. It is ineffective to say "There were three reasons why we modified the test rig in this way. One was to . . .", meanwhile counting them off using hand-waving body language, with nothing on the screen.

Instead, prepare a visual aid that lists it in point form, and expand on each one while you speak. It is very important not to read the slide woodenly word for word.

Example of design of slide
Reasons for modifying the test rig:
1. xxx
2. xxx
3. xxx

Possible spoken material
There were three reasons why we modified the test rig.
The first was that we found that . . .
We also found that . . . so we introduced a widget that would . . .
The third reason was that . . .

Having the key points in front of you on the display also avoids your having to use written notes.

Common mistakes
- Insufficient planning of your visual material to ensure that what you are saying appears simultaneously on the screen in point form
- To put uninformative points on the screen, while simultaneously overwhelming the audience with detailed data

18.7.2 Make sure that the audience can read your text and see your illustrations

Use a large type size—probably much larger than is instinctive.
At least 25-point-type is needed for text, even for a small room, and correspondingly larger for lecture theaters. Make sure that it is big enough to be easily read from the back of the room.

Diagrams should be large and chunky and should have large lettering.
Do not just take unaltered diagrams out of a document; they will be far too small and the lines will be too thin. Make them large and thicken the elements. Also make sure that the lettering is thick, particularly on the axes and legends of graphs.

Tables—if you really need them—should be very simple.
It is not unusual in conferences or seminars to see tables, probably copied from a journal paper, that are full of minute text that no one can see. Tables are best avoided for talks; try instead using other methods of data presentation. If you think you must use a table, take the time to redesign it by rigorous selection and deletion so that it is easily seen and understood.

18.7.3 Use an uncluttered layout: Don't put too much on one overhead or slide

Crowded visual aids are fussy and difficult to read. They should be made up of only a few key points.

18.7.4 Don't write complete sentences on your visual aids

It is too much for the audience to read. Most of the text should be in point form.

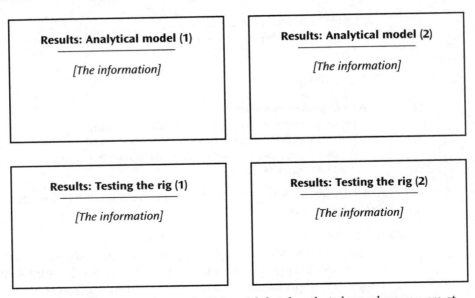

Figure 18-4 *How to keep the audience oriented: slides with headers that show where you are at any one point within your framework*

18.7.5 Dealing with detail: Keep orienting your audience to the framework

Make sure that your visual aids show clearly where you are in the scheme of the talk. When you need to present detailed material, ensure that your visual aid shows not only the detail but also some pointers to the general framework.

This can be done by using headers at the top of your slides so that the audience can see where you are within the structural framework (Figure 18-4). If an audience member's attention wanders slightly, as often happens, it is then simple to reorient to the framework.

18.7.6 Proofread your visual aids for spelling

Use both computer and human resources, and do it immaculately. There will be people in the audience who are expert spellers, and errors are very unprofessional.

18.7.7 Be aware of how to use presentation software effectively

Presentation software such as Microsoft PowerPoint® or similar presentation software can produce superb visual material if handled intelligently. Here are a number of traps.

What looks good on a monitor may be poor when projected.

Colors and brightness can often appear more intense on a monitor. It is essential to project your material to judge the final effect. Particular problems arise from color intensity and hue, and the clarity of material on a nonhomogeneous background (e.g., graded color or a patterned or photographic background).

Be very careful in your choice of background.

Some of the presentation software standard backgrounds are not well designed. When words and diagrams are superimposed on them, the final effect can be messy. Choose a simple, uncluttered background. Alternatively, design your own, keeping it simple and unfussy. Many participants will be familiar with the standard backgrounds; an effective self-designed background brings life to your presentation.

Be careful in your choice of colors.

Make sure that the color combinations do not obscure detail or make it difficult to read the words. Some color combinations are very poor for showing detail.

- ► Good combinations:
  - White/clear yellow on black/dark blue (not light blue)
  - Black/dark blue on white
- ► Particularly problematic:
  - Avoid using red; it does not have good resolution. For example, in a graph with several colored lines, the red line will look unclear.

Don't be tempted to go to extremes with the animation functions.

Flying bullets and dissolving transitions can be beguiling at first sight, but they may irritate an expert audience.

18.8   Delivering your presentation

Overview: Delivering your talk
- Your voice
- The position in which you stand
- The gestures you make
- Interaction with visual aids
- Eye contact
- Pointing
- Losing your place or having to pause
- Interruptions
- Timing
- Needing to finish in a hurry
- Get to know the venue

18.8.1   Your voice

Make your voice louder and slightly more deliberate than your normal speaking voice unless there is good amplification. This needs conscious control. It is not easy to do, because you always feel that you sound peculiar.

If you have the courage, try recording yourself. You probably will be shocked when you hear yourself—most people are.

Be aware of verbal tics. A recording will let you know if you repeatedly and unconsciously use certain words. Favorites are: basically, you know, sort of, like, uuuum, and anduuh. Try to avoid them.

Common mistakes with your voice when you are nervous

- Your voice speeds up.
- Your voice becomes monotonous. Try hard to avoid this; it results in a boring presentation.
- You think your voice is cracking or wobbling. Most of the time it will not be apparent to the audience. Don't worry about it.
- Your voice is quieter than usual.
- *For nonnative speakers of English:* The problem of a quiet, rapid voice is a common one, either because of (1) nervousness about making grammatical mistakes, or (2) feeling that you are sufficiently fluent but forgetting that you may slur words together or have a strong accent. Slow down, and try to speak deliberately and positively, and more loudly. Do not worry about grammatical mistakes. It is better to say something ungrammatical in a strong voice—your audience will understand and be sympathetic.

18.8.2   The position in which you stand

Be conscious of whether you are blocking the view of the screen for part of the audience. If this is unavoidable, try to move around so that the view is blocked only intermittently.

18.8.3   The gestures you make

Try to add gestures to your speaking, but move naturally. Most people use their hands when they speak privately; however, nervousness can cause rigidity, overextravagant gestures, or a demeanor that is too informal.

What to do with hands

Many people find that they suddenly become very conscious of their hands and do not know what to do with them. There is no easy answer to this. Clasping them behind your back is sometimes suggested, but it is not an ideal solution.

Common mistakes with hands

- Putting them in pockets. This looks unprofessional.

- Putting them on your hips or hooking the thumbs into the waistband or pockets. This may look aggressive.
- Fiddling with a pen, the pointer, or something in your pocket.

18.8.4   Interaction with visual aids

It makes for a bleak presentation if you just put up your visual aids and don't interact with them. Leaving your place to point out a feature of interest livens it up. But in the stress of being center stage, there are a few traps to be aware of.

Common mistakes

- *Leaving something on the screen long after you have finished talking about it.* Your visual aids should be planned so that whatever you are saying is backed up by what is on the screen.
- *Leaving the audience to navigate their way through a complicated figure.* It is not good to show, for example, a complex diagram and say something like "As you can see, angle α is much smaller than angle . . ." and leaving the audience to find it themselves. Point it out.
- *If you are using an overhead projector, muddling the overhead slides.* Make sure you manipulate the slides logically. As you take each one off the projector, place it in a separate stack of the used ones. You need to make sure that your used stack stays separate from the ones that are still to come. Piling them all on top of each other often means you have the embarrassment of having to shuffle through them to find the right one. The used stack also needs to remain in order so that they can be easily found again if you need them at question time.
- *If you are using an overhead projector, using the striptease system of displaying a slide* (sequentially revealing more and more by sliding a piece of paper down it). It is visually unappealing and may be irritating. There is no harm at all in allowing the audience to see the rest of the slide before you deal with it.

18.8.5   Eye contact

Texts on public speaking emphasize the importance of looking at people in the audience and making each member feel that you are personally speaking to him or her. This, of course, is easy advice to give, but it is terrifying to put into practice.

Advice if you are nervous: Try to make yourself look up at the audience as much as possible. This gives a much better impression than just reading, but it is easier said than done.

Advice if you are more than a beginner: Looking people in the eye is difficult. However, if you scan across the audience at chin level, you will avoid people's direct gaze and it will be undetectable except to the people in the front row. Don't scan above the heads of the people in the back row. Although doing so is sometimes advised, it is poor advice because you will appear completely out of contact.

18.8.6 Pointing

You can point at either the screen or directly at the overhead projector glass.

For the screen, use the wooden pointer that is almost always a staple feature of lecture theaters or a laser pointer, which may be a problem if your hand is shaking.

Common mistakes when using the screen
- Becoming too absorbed in the material on the screen and losing contact with the audience.
- Pointing with your finger from a distance at something on the screen and forgetting that your line of sight is different from that of the audience. You can see it at your fingertip, but the audience won't know what you're pointing at.
- Using a laser pointer too frenetically; rapidly circling an item of information instead of directing a steady beam at it.

For the overhead projector glass, use a pen or pencil with a tip to it. If you are nervous and the pen is shaking, place it on the glass. Avoid one that is circular in cross section—it could roll.

Do not use a finger. It is too blunt and sticky (the foil may move), and it looks unprofessional.

Common mistakes when using the overhead projector glass
- Blocking the view of a section of an audience.
- Standing rigidly. If you use the screen, it will force you to move from the projector to the screen and back again, which makes you look more dynamic.
- Crouching over the projector and ignoring the audience.
- Fiddling with the overhead slide and frequently causing it to twitch.
- When using a small pointer such as a pen to point at the overhead projector glass:
  — Moving the pen around too fast
  — A visibly shaking pen
  — A rolling pen

18.8.7 Losing your place or having to pause

Losing your place or having to pause for any reason is unnerving. To you, the pause can seem embarrassingly long and obvious. But it is worth remembering that what seems endless to you will not be noticed by the audience as long as you don't draw attention to it by becoming flustered and muttering inanities.

The best strategy is to say nothing, control your body language, and as calmly as possible find your place again or collect your thoughts.

18.8.8   Interruptions

If there is an interruption beyond your control and you find your audience's attention attracted away from you, just say "I'll repeat that" without flustering. It brings the control back to you.

18.8.9   Timing

Many people run overtime, even when they have extensively practiced the presentation. A number of factors can contribute to this.

A common major problem is to find that in the stress of the moment, you deviate from your planned script and include something that you did not intend to say. If this happens a few times, you will run seriously overtime.

When you practice, you need to speak out loud. When you read to yourself or whisper, you will take less time than you would when you present.

When you practice, make sure you also change the visual aids, point at the screen, and so on. This adds to the time needed. Novices often comment that they had not allowed for this in their timing practice.

18.8.10   Needing to finish in a hurry

If you run out of time and the chairperson asks you to finish up, do the following:

▶ Without fluster, put up your Conclusions slide and say something like "I'm sorry—I haven't the time to finish this section. If anyone needs the detail, I'll be glad to talk to them afterwards. To finish, these are the main conclusions from our work." If you think you can take a few more seconds, you could very briefly talk about them. If not, leave them there for the audience to read while you take your leave.

▶ If you are using presentation software, make sure that the Conclusions slide is the very last slide in the series. Then all that is necessary is to hit the End key, and it will go straight to it. If, for your complete presentation, you plan to show slides after the Conclusions slide (for example, an Acknowledgments slide), copy the Conclusions slide so that it is also in the final position. This is by far the most professional way to finish in a hurry, and it emphasizes the importance of having an effective Conclusions slide.

18.8.11   Get to know the venue

Make sure that you find the time to visit the venue and check it out. It is vital to find out beforehand the positions of the various controls for the visual aid equipment and the lights. It is unnerving to try to locate a control during your talk and to ask for help.

18.9   Answering questions

The thought of answering questions unnerves many people, particularly non-native speakers. Remember, though, that a good chairperson should help clarify questions.

▶ **Work out beforehand the questions that you might be asked.**

Try to visualize your presentation through the audience's eyes. You are unlikely to be taken by surprise if you have prepared well.

▶ **Prepare a supplementary group of PowerPoint slides or overhead slides that are more detailed than those you used in the talk.**

They could perhaps show more detailed data or greater detail of your procedures. The type size does not need to be as large as that on your main visual aids. You can often use one of these to answer a question very efficiently.

▶ **Make sure that you understand the question correctly.**

This can be particularly problematic and worrisome for nonnative speakers of English. Do not answer a question until you are sure you understand it.

> *Suggested wording*
> I'm sorry; I didn't understand that. Could you repeat it, please?

Be prepared for the questioner who gives a long discourse in which the main point can often be buried. If you cannot understand where the question lies, ask the questioner to clarify it. If it is still obscure, the chairperson should help.

▶ **If you still don't understand the question, don't be afraid to ask for further clarification.**

It is better to ask for repeats than to be flustered into answering wrongly.

> *Suggested wording*
> I'm sorry, I still didn't understand. Could you clarify it for me, please?

After this point the onus is on the questioner, and the chairperson should help to clarify it.

▶ **If you think that the question has not been heard by the rest of the audience: Repeat it yourself.**

Say, "The question was 'How does the ...?' " and then answer it.

▶ **If you do not know the answer to a question, be honest.**

Don't try to fudge your way through an answer and hope that the audience doesn't notice that you are trying to cover up. It is always obvious when a speaker is doing this.

Either say in a positive voice that you don't know or offer to find out the answer. This is a clear indication of honesty and willingness to communicate the material and that you are confident in your work.

Suggested wording

I don't know the answer to that question, I'm afraid.

I don't have the answer to that at the moment, but I'll find out for you by tomorrow.

▶ **Be honest about your problems, but not negative.**

Don't be afraid to mention—briefly, objectively, and without emotion—any difficulties you may have had with your work. It is not a sign of weakness; everyone in the audience will be able to relate to it, and someone may be able to help. But make sure that you do not present yourself as self-pitying.

Checklist for an oral presentation

Planning

☐ Have you been extremely selective and concentrated on your main points?

☐ Are you aware of what people can do when they are nervous, and have you planned to avoid such behaviors?

☐ Have you planned the whole talk and each section of the talk as a diamond by starting and ending with overview information?

☐ Have you planned your visual aids so that each key point will be simultaneously spoken and displayed in point form on the screen?

☐ Do you know the first couple of minutes of your presentation by heart?

☐ Do you have a title slide, giving your name, institution, and the title of the talk?

☐ Does your second slide give an overview of the whole talk? Does it give meaningful information?

☐ Do you have a final slide to sum up your conclusions?

☐ Are you using notes? Could you possibly do without notes and take your cues from the audiovisual material?

Notes

☐ Are they in spoken style rather than a formal written style?

☐ Is the printing large enough for you to be able to read them easily while under pressure?

☐ If you need a full script, can you work from an annotated script using keywords?

☐ If you lose your place, can you readily find your way back into your notes while under pressure?

The presentation

☐ Have you planned how you are going to greet the audience?

If you are using overhead slides:

☐ Are they in order?

☐ Have you removed the backing sheets?

☐ Do you know where the on/off switch of the overhead projector is?

☐ If you are using presentation software with a projector: Have you checked the equipment at the venue?

☐ Is there a clear framework for the audience to be able to orient themselves to your material? Is it visible on your visual aids? Will the audience be constantly aware of where you are within the framework?

☐ Are you using the style of spoken English, not written English?

☐ Are you avoiding pompous English? Are you speaking to inform, not to impress?

☐ Have you planned to look up as much as possible?

☐ Is your final Conclusions slide at hand for use if you have to finish in a hurry?

Visual aids

☐ *Important:* Will each key point be simultaneously spoken and backed up in point form on the screen?

☐ Is the type size at least 25 points?

☐ Have you avoided cramming too much onto the visual aids?

☐ Are the visual aids simply designed?

☐ Are the illustrations large, chunky, and easily visible?

☐ Is the use of color and pattern simple and effective?

Answering questions

☐ Have you worked out beforehand the possible questions?

☐ Do you have a supplementary set of more detailed visual aids that you can use for answering questions?

☐ Are you prepared for not understanding a question?

19

A Presentation to a Small Group

This chapter gives guidelines for a presentation to a much smaller audience than that at a seminar or conference.

Structure of the chapter

19.1   The constraints of presenting to a small group

The meeting is likely to take place across a table, with no means of projecting visual aids.

19.2   Basic principles for preparation

- ▶ Visualize yourself and your material through the audience's eyes.
- ▶ Work out beforehand the questions you may be asked.
- ▶ Identify the main points of the work and its strengths.
- ▶ Identify the key weak points and problems, and prepare yourself for questions about them.
- ▶ Think graphically. Clear graphical visual aids are an effective means of making points and answering questions.

19.3   A professional interview or an oral examination

- ▶ Prepare a summary of the work that you have done and its significance.
- ▶ Keep in mind what you have done, how you have done it, and what is new and significant about it.
- ▶ After the initial small talk to make you feel more comfortable, many oral examinations start with a request to the candidate to summarize his or her work. To prepare for this beforehand, you need to be able to stand back from the minute details and prepare an overview.
- ▶ Make sure that you can answer the following possible questions:
  - What is the significance of your work?
  - What skills did you develop?
  - If you were to do it again, would you approach it differently?
  - Where do you see it leading?
- ▶ If you have submitted a large body of written work, make sure you can navigate your way around it without stumbling. You may need to refer to it to answer questions.
- ▶ During the examination or interview, if you are asked a question that needs deliberation, allow yourself time to think without getting flustered. Don't let the pressures of the moment force you into a hasty answer. Your assessors will prefer a period of thought followed by a reasoned answer to one that is unconsidered or hasty.

19.4   A presentation to a review panel

An example of this type of presentation is a progress report to another organization.

19.4.1   Visualize yourself through the audience's eyes

You want the audience to listen to your message, understand it, and be influenced by it. Keep in mind:
- The particular concerns of the individuals in the panel (commercial, technical, etc.).
- They may not have very much prior knowledge of your work. What is obvious to you may not be so to them.
- The significance to them of each point you make, for example impact on part numbers, costs, and assembly time without reducing the quality of the outcome.
- Summarize the take-home message. It can be couched in terms of economic feasibility, fixed and variable cost savings, projected break-even points, payback period, and so on.

- The possible barriers to getting your ideas accepted need to be identified.
- Most people prefer concrete examples, not concepts.

19.4.2   Identify the key points and plan a flip chart

▶ Be rigorously selective in what you will present. There is never enough time to say everything.

▶ Aim to present all your top material in the first few minutes. Use the same diamond structure as for a report or conference presentation:
  - Initial overview of the main points
  - The main body
  - The summing up

▶ Identify the take-home message—the main point—and don't be afraid to present it three times. Include it in your initial overview, the main body, and the final summing up.

▶ Collect the information into a flip chart. You then have just one folder, not an array of drawings that can become muddled. Necessary information in the flip chart:
  - The first item should be your initial overview slide containing the main points.
  - A graphic that summarizes your approach to the project should be included.
  - If you are recommending solutions, make sure that the key recommendations are clearly shown. Don't overwhelm people with a large number of recommendations—prioritize, stating the important features.
  - An ongoing presentation that shows the current status of the project so that it can be presented at a moment's notice. Don't be afraid of not having everything 100% correct.

19.4.3   Think graphically. There is no substitute for good graphics.

The quality of the graphics is vital. Scientists and engineers think graphically. Time can be saved by using presentation software such as Microsoft Power-Point.

You don't need many graphics, perhaps a few outstanding drawings and slides.

If you are using existing graphics, add value to what you already have and take away the clutter (use whiteout solution and a photocopier). For example, you may not need the dimensions for an overview diagram.

Checklist presentations to small groups

For a presentation to a small panel

☐ Did you visualize yourself and your material through the audience's eyes?

☐ Did you work out beforehand the questions you may be asked?

☐ Did you identify the main points of the work and its strengths?

☐ Did you identify the key weak points and problems, and prepare yourself for questions about them?

☐ Did you think graphically? Clear graphics are an effective means of making points and answering questions.

For an oral examination

☐ Have you allowed yourself adequate time to become thoroughly familiar with the material and the thesis again?

☐ Have you prepared a summary of the work that you have done?

☐ Have you thought about the significance of your work: what you have done, how you have done it, and what is new about your research?

☐ Are you sure you can navigate your way around your thesis without hesitation?

☐ Have you recently done a literature search for any new work that may have come out since you submitted your thesis?

Can you answer the following questions:

    ☐ What is the significance of your work?

    ☐ What skills did you develop?

    ☐ If you were to do it again, would you approach it differently?

    ☐ Where do you see it leading?

☐ Have you worked out other questions you might be asked?

☐ Can you identify the main points of the work and its strengths?

☐ Do you know its weak points, and are you prepared for questions on them?

For a review panel

☐ Do you know the particular concerns of the individuals in the panel (commercial, academic, etc.)?

☐ Do you know how much prior knowledge of your work they have?

☐ Can you gauge the significance to them of each point you make?

☐ Can you summarize the take-home message of your work?

☐ Can you couch it in terms that are meaningful to the panel (economic feasibility, fixed and variable cost savings, projected break-even points, payback period, etc.)?

☐ Can you identify any possible barriers to getting your ideas accepted?

☐ Have you been rigorously selective in what you will present?

☐ Can you present all your top material in the first few minutes?

□ Have you prepared an initial overview of the main points and a final summing up?

□ Will you present the take-home message—the main point—three times: initial overview, the main body, and final summing up?

□ Are you using a flip chart?

Have you included the following information:

    □ An initial overview slide containing the main points?

    □ A graphic that summarizes your approach to the project?

    □ Clearly demonstrated key recommendations?

    □ An ongoing presentation that shows the current status of the project?

□ Is the quality of your graphics very good?

□ Have you reduced the clutter of existing graphics?

References and Resources

Part 7 gives suggestions for further resources on engineering writing.

Both hard copy and electronic sources are given. However, electronic sources can sometimes disappear over time, either by alteration of the URL or by removal of the site. If this happens, it is recommended that search terms be used with an effective search engine or electronic database to trace either the original site or similar material.

References
and Resources

General engineering writing

Beer, David F. (ed.) (1991). *Writing & Speaking in Technology Professions: A Practical Guide.* New York: Wiley-IEEE Computer Society Press.

Blicq, Ronald S., and Moretto, Lisa (2003). *Technically-Write!* 6th Edition. Upper Saddle River, N.J.: Prentice Hall.

Haines, Roger W., and Bahnfleth, Donald R. (1990). *Roger Haines on Report Writing: A Guide for Engineers,* 1st Edition. Blue Ridge Summit, Pa.: Tab Professional & Reference.

Hargis, Gretchen (ed.) (1998). *Developing Quality Technical Information: A Handbook for Writers and Editors.* Upper Saddle River, N.J.: Prentice Hall.

Moretto, Lisa, and Blicq, Ronald S. (1995). *Writing Reports to Get Results: Quick, Effective Results Using the Pyramid Method,* 2nd Edition. New York: Wiley-IEEE Computer Society Press.

Nagle, Joan G. (1996). *Handbook for Preparing Engineering Documents: From Concept to Completion.* New York: IEEE Press.

Piotrowski, Maryann V. (1996). *Effective Business Writing: A Guide for Those Who Write on the Job,* 2nd Edition. New York: HarperPerennial.

Rigby, David W. (2001). *Technical Document Basics for Engineering Technicians and Technologists.* Upper Saddle River, N.J.: Prentice Hall.

Sides, Charles H. (1999). *How to Write and Present Technical Information,* 3rd Edition. Cambridge, U.K.: Cambridge University Press.

Silyn-Roberts, Heather (2000). *Writing for Science and Engineering: Papers, Presentations and Reports.* Oxford, U.K.: Butterworth-Heinemann.

Proposals

Bartlett, Robert R. (1997). *Preparing International Proposals*. Reston, Va.: American Society of Civil Engineers.

Friedland, Andrew J., and Folt, Carol L. (2000). *Writing Successful Science Proposals*. New Haven, Conn.: Yale University Press.

Specifications

Abdallah, Eli T. (ed.) (2000). *Preparing Specifications for Design-Bid Build Projects*. Reston, Va.: American Society of Civil Engineers.

Architectural Products Specifications Guide. Sherwin-Williams Company. http://www2.sherwin-williams.com/architects/specguide/.

Cox, Peter J. (1994). *Writing Specifications for Construction*. London: McGraw-Hill.

Developing Specifications for Purchasing. Department of Public Works, Queensland Government, Australia. http://www.qgm.qld.gov.au/bpguides/specif/.

Oriel, John. *Guide to Specification Writing for U.S. Government Engineers*. http://www.ntsc.navy.mil/Resources/Library/Acqguide/spec.htm.

Purdy, David C. (1991). *A Guide to Writing Successful Engineering Specifications*. New York: McGraw-Hill.

Rosen, Harold J. (1999). *Construction Specifications Writing: Principles and Procedures*, 4th Edition. New York: John Wiley & Sons.

Specification Writing Best Practice Advice. Victorian Government Purchasing Board, Australia. http://www.vgpb.vic.gov.au/CA256C450016850B/0/75C6E5ADDC457DB4CA256C7E00819769?OpenDocument.

Willis, Christopher J., and Willis, Andrew J. (1997). *Specification Writing for Architects and Surveyors*, 11th Edition. Malden, Mass., Blackwell Science.

Manuals and procedures

Cloud, Phillip A. (2001). *Developing and Managing Engineering Procedures: Concepts and Applications*. Park Ridge, N.J.: Noyes Publications.

Foshay, Wellesley R. (2003). *Writing Training Materials That Work: How to Train Anyone to Do Anything*. San Francisco, Calif.: Jossey-Bass/Peiffer. (Book and CD-ROM edition.)

Haydon, Leslie M. (1995). *The Complete Guide to Writing and Producing Technical Manuals*. New York: John Wiley & Sons.

Horton, William K. (1994). *Designing and Writing Online Documentation: Hypermedia for Self-Supporting Products*, 2nd Edition. New York: John Wiley & Sons.

Kurtus, Ron. *The Process of Writing a Technical Manual.* Technologies and the School for Champions. http://www.school-for-champions.com/techwriting/ techprocess.htm.

Peabody, Larry, and Gear, John (1998). *How to Write Policies, Procedures and Task Outlines,* 2nd Edition. Ravensdale, Wash.: Idyll Arbor.

Thirlway, Martyn (1994). *Writing Software Manuals: A Practical Guide.* New York: Prentice Hall.

Tidwell, Mike (2000). *How to Produce Effective Operations and Maintenance Manuals.* Reston, Va.: American Society of Civil Engineers.

Tognazzini, Bruce. *How to Publish a Great User Manual.* http://www.asktog. com/columns/017ManualWriting.html.

Letters

Bond, Alan (1998). *Over 300 Successful Business Letters for All Occasions.* Hauppage, N.Y.: Barrons Educational Series.

Danziger, Elizabeth (2001). *Get to the Point! Painless Advice for Writing Memos, Letters and E-mails Your Colleagues and Clients Will Understand.* New York: Three Rivers Press.

Geffner, Andrea (1998). *Business English: A Complete Guide to Developing an Effective Business Writing Style,* 3rd Edition. Hauppage, N.Y.: Barrons Educational Series.

Griffin, Jack (1997). *The Complete Handbook of Model Business Letters.* Upper Saddle River, N.J.: Prentice Hall.

Publicity material

Franklin, Reece (1996). *The Consultant's Guide to Publicity: How to Make a Name for Yourself by Promoting Your Expertise.* New York: John Wiley & Sons.

Marconi, Joe (1999). *The Complete Guide to Publicity.* New York: McGraw-Hill.

Yale, David R., and Carothers, Andrew J. (2001). *The Publicity Handbook, New Edition: The Inside Scoop from More than 100 Journalists and PR Pros on How to Get Great Publicity Coverage,* 2nd Edition. New York: McGraw-Hill.

Yale, David R., and Knudsen, Anne (eds.) (1995). *Publicity & Media Relations Checklists.* New York: McGraw-Hill.

Journal papers

Booth, Vernon (1993). *Communicating in Science: Writing a Scientific Paper and Speaking at Scientific Meetings,* 2nd Edition. Cambridge, U.K.: Cambridge University Press.

Carter, Sylvester P. (1987). *Writing for Your Peers: The Primary Journal Paper*. New York: Praeger Publishers.

Davis, Martha (1996). *Scientific Papers and Presentations*. San Diego, Calif: Academic Press.

Day, Robert A. (1998). *How To Write & Publish a Scientific Paper*, 5th Edition. Phoenix, Ariz.: Oryx Press.

Introduction to Journal-Style Scientific Writing. Department of Biology, Bates College, Lewiston. http://abacus.bates.edu/~ganderso/biology/resources/writing/HTWgeneral.html.

Periodical Title Abbreviations (2001). Volumes 1–2. Detroit, Mich.: Gale Research Company. (Volume 1 enables search by abbreviation, Volume 2 by title.)

Oral presentation

Beer, David F. (ed.) (1991). *Writing & Speaking in Technology Professions: A Practical Guide*. New York: Wiley-IEEE Computer Society Press.

Booth, Vernon (1993). *Communicating in Science: Writing a Scientific Paper and Speaking at Scientific Meetings*, 2nd Edition. Cambridge, U.K.: Cambridge University Press.

Boylan, Bob (2001). *What's Your Point? The 3-Step Method for Making Effective Presentations*. Holbrook, Mass.: Adams Media Corporation.

Leech, Thomas (1993). *How to Prepare, Stage and Deliver Winning Presentations*, 2nd Edition. New York: AMACOM.

Mablekos, Carole M. (1991). *Presentations That Work* (IEEE Engineers Guide to Business, Vol. 1). New York: IEEE.

Zelazny, Gene (1999). *Say It with Presentations: How to Design and Deliver Successful Business Presentations*. New York: McGraw-Hill.

Conference poster

Briscoe, Mary, H. (1997). *Preparing Scientific Illustrations: A Guide to Better Posters, Presentations, and Publications*, 2nd Edition. New York: Springer Verlag.

Gray, Norman, and Stewart, Graeme. (2003). *Tutorial: Using LaTeX to Produce Conference Posters*. http://www.astro.gla.ac.uk/users/norman/docs/posters/.

National Arts Centre, Ottawa. *How to Create a Conference Poster*. http://www.mitacs.math.ca/AC2003/index.php?section=poster.

Scott, Rich. *Tutorial: PowerPoint—Creating Posters*. http://www.cmer.wsu.edu/~yonge/ce465/poster.pdf.

Style of writing

Burchfield, R. W. (ed.), and Fowler, Henry W. (2000). *The New Fowler's Modern English Usage*, 3rd Edition. New York: Oxford University Press.

Cappon, Rene J., and Cappon, Jack (2003). *The Associated Press Guide to Punctuation.* Cambridge, Mass.: Perseus Publishing.

Gowers, Ernest, Greenbaum, Sidney, and Whitcut, Janet (2003). *The Complete Plain Words*. Boston, Mass.: David R. Godine.

Grossman, John (2003). *The Chicago Manual of Style: The Essential Guide for Writers, Editors, and Publishers*, 15th Edition. Chicago: University of Chicago Press.

Strunk, William, Jr. (2000). *The Elements of Style,* 4th Edition. Boston: Allyn and Bacon.

Truss, Lynne (2004). *Eats, Shoots & Leaves: The Zero Tolerance Approach to Punctuation*. New York: Gotham Books.

Quick Reference Guide

The Parts of Speech and Verb Forms

Parts of speech

Verbs
- Words that indicate action: what is done, what was done, or what is said to be.

  The ship <u>sailed</u>.

Nouns
- Names, things.

  <u>Columbus</u> sailed in the <u>ship</u>.

Pronouns
- Words used instead of nouns so that nouns need not be repeated.

  <u>He</u> sailed in <u>it</u>.

Adjectives
- Words that describe or qualify nouns.

  The <u>tall</u> man sailed in the <u>big</u> ship.

Adverbs
- Words that modify verbs, adjectives, and other adverbs. They often end in –ly.

  The big ship <u>slowly</u> sailed past the <u>steeply</u> sloping cliffs.

Prepositions
- Each preposition marks the relationship between a noun or pronoun and some other word in the sentence.

  The ship sailed <u>past</u> the cliffs and <u>across</u> the sea <u>to</u> America.

Conjunctions

- Words used to join the parts of a sentence or to make two sentences into one: and, but, so, because, as, since, while.

The ship sailed to America <u>and</u> came straight back.

The ship sailed to America <u>but</u> did not stay long.

The ship sailed fast, <u>so</u> it got there quickly.

The ship sailed slowly <u>because</u> *(or* as *or* since*)* the sails were torn.

The people on the dock waved <u>while</u> the ship sailed away.

Gerund

- A word ending in –ing that behaves in some ways like a noun and in some ways like a verb.

She likes <u>using</u> a computer.

You can save electricity by <u>switching</u> off the lights.

Tenses and forms of the verb

This section describes, in very simple terms, the various forms of a verb.

Present

- Describes what is happening at the moment:

The sun <u>shines</u>.
Chlorofluorocarbons <u>cause</u> ozone depletion.

The sun <u>is shining.</u>
Global warming <u>is developing</u> into a major environmental issue.

Past

- Describes what happened in the past.

The sun <u>shone</u>.
The burning of fossil fuels <u>caused</u> carbon dioxide levels to rise.

The sun <u>was shining</u>.
By the end of the twentieth century, carbon dioxide levels <u>were causing</u> temperature levels to rise.

The sun <u>has shone</u>.
The burning of fossil fuels <u>has caused</u> levels of carbon dioxide to rise.

The sun <u>has been shining.</u>
Carbon dioxide levels <u>have been causing</u> concern for a long time.

The sun <u>had shone.</u>
By the 1950s, carbon dioxide levels in the atmosphere <u>had risen</u> to 315 ppm.

The sun <u>had been shining.</u>
By the end of the twentieth century, carbon dioxide levels <u>had been rising</u> for a number of decades.

Future

- Describes what is going to happen in the future.

  The sun <u>will shine</u>.
  Increased emission of greenhouse gases <u>will cause</u> a change in the global climate.

  The sun <u>will be shining</u>.
  By the middle of this century, the increased emission of greenhouse gases <u>will be causing</u> a global change in climate.

  The sun <u>will have shone</u>.
  By the middle of this century, carbon dioxide levels <u>will have risen</u> to twice the preindustrial level.

  The sun <u>will have been shining.</u>
  By the middle of this century, carbon dioxide levels <u>will have been causing</u> concern for many decades.

Conditional

- Expresses a condition. Sometimes needed in recommendations.

  This result could imply that… .

  The test equipment should be modified as shown.

Subjunctive

- In technical writing, usually used in recommendations.

  It is recommended that the system <u>be</u> upgraded.

  It is recommended that the manager <u>assess</u> the effects of the change.

Imperative

- The form of the verb that gives an instruction. The preferred form of the verb for a procedure or set of instructions. Sometimes used in recommendations.

  Turn the power off.

  Do not open the valve before it cools to 18°C.

  Automate the operation of the sluice gates.

Index

Sections of a document have been capitalized and italicized, e.g., *Abstract, List of References*, and so on. Types of documents have been capitalized but not italicized, e.g., Due Diligence Report, Feasibility Study, and so on.